RADON IN THE ENVIRONMENT

Studies in Environmental Science

Other volumes in this series

1 **Atmospheric Pollution 1978** edited by M.M. Benarie
2 **Air Pollution Reference Measurement Methods and Systems**
 edited by T. Schneider, H.W. de Koning and L.J. Brasser
3 **Biogeochemical Cycling of Mineral-Forming Elements**
 edited by P.A. Trudinger and D.J. Swaine
4 **Potential Industrial Carcinogens and Mutagens** by L. Fishbein
5 **Industrial Waste Management** by S.E. Jørgensen
6 **Trade and Environment: A Theoretical Enquiry** by H. Siebert, J. Eichberger, R. Gronych and R. Pethig
7 **Field Worker Exposure during Pesticide Application** edited by W.F. Tordoir and E.A.H. van Heemstra-Lequin
8 **Atmospheric Pollution 1980** edited by M.M. Benarie
9 **Energetics and Technology of Biological Elimination of Wastes**
 edited by G. Milazzo
10 **Bioengineering, Thermal Physiology and Comfort** edited by K. Cena and J.A. Clark
11 **Atmospheric Chemistry. Fundamental Aspects** by E. Mészáros
12 **Water Supply and Health** edited by H. van Lelyveld and B.C.J. Zoeteman
13 **Man under Vibration. Suffering and Protection** edited by G. Bianchi, K.V. Frolov and A. Oledzki
14 **Principles of Environmental Science and Technology** by S.E. Jørgensen and I. Johnsen
15 **Disposal of Radioactive Wastes** by Z. Dlouhý
16 **Mankind and Energy** edited by A. Blanc-Lapierre
17 **Quality of Groundwater** edited by W. van Duijvenbooden, P. Glasbergen and H. van Lelyveld
18 **Education and Safe Handling in Pesticide Application** edited by E.A.H. van Heemstra-Lequin and W.F. Tordoir
19 **Physicochemical Methods for Water and Wastewater Treatment** edited by L. Pawlowski
20 **Atmospheric Pollution 1982** edited by M.M. Benarie
21 **Air Pollution by Nitrogen Oxides** edited by T. Schneider and L. Grant
22 **Environmental Radioanalysis** by H.A. Das, A. Faanhof and H.A. van der Sloot
23 **Chemistry for Protection of the Environment** edited by L. Pawlowski, A.J. Verdier and W.J. Lacy
24 **Determination and Assessment of Pesticide Exposure** edited by M. Siewierski
25 **The Biosphere: Problems and Solutions** edited by T.N. Veziroğlu
26 **Chemical Events in the Atmosphere and their Impact on the Environment** edited by G.B. Marini-Bettòlo
27 **Fluoride Research 1985** edited by H. Tsunoda and Ming-Ho Yu
28 **Algal Biofouling** edited by L.V. Evans and K.D. Hoagland
29 **Chemistry for Protection of the Environment 1985** edited by L. Pawlowski, G. Alaerts and W.J. Lacy
30 **Acidification and its Policy Implications** edited by T. Schneider
31 **Teratogens: Chemicals which Cause Birth Defects** edited by V. Kolb Meyers
32 **Pesticide Chemistry** by G. Matolcsy, M. Nádasy and V. Andriska
33 **Principles of Environmental Science and Technology (second revised edition)** by S.E. Jørgensen
34 **Chemistry for Protection of the Environment 1987** edited by L. Pawlowski, E. Mentasti, C. Sarzanini and W.J. Lacy
35 **Atmospheric Ozone Research and its Policy Implications** edited by T. Schneider, S.D. Lee, G.J.R. Wolters and L.D. Grant
36 **Valuation Methods and Policy Making in Environmental Economics** edited by H. Folmer and E. van Ierland
37 **Asbestos in the Natural Environment** by H. Schreier
38 **How to Conquer Air Pollution. A Japanese Experience** edited by H. Nishimura
39 **Aquatic Bioenvironmental Studies: The Hanford Experience, 1944–1984** by C.D. Becker

Studies in Environmental Science 40

RADON IN THE ENVIRONMENT

M. WILKENING

Department of Physics, New Mexico Institute of Mining and Technology, Socorro, NM 87801, U.S.A.

ELSEVIER
Amsterdam — Oxford — New York — Tokyo   1990

ELSEVIER SCIENCE PUBLISHERS B.V.
Sara Burgerhartstraat 25
P.O. Box 211, 1000 AE Amsterdam, The Netherlands

Distributors for the United States and Canada:

ELSEVIER SCIENCE PUBLISHING COMPANY INC.
655, Avenue of the Americas
New York, NY 10010, U.S.A.

```
Library of Congress Cataloging-in-Publication Data

Wilkening, M. (Marvin)
    Radon in the environment / M. Wilkening.
      p.   cm. -- (Studies in environmental science ; 40)
    Includes bibliographical references and index.
    ISBN 0-444-88163-8
    1. Atmospheric radon--Environmental aspects. 2. Atmospheric
  radon. 3. Atmospheric diffusion--Research.  I. Title. II. Series.
  TF885.5.R33W55  1990
  628.5'35--dc20                                          90-3779
                                                             CIP
```

ISBN 0-444-88163-8

© Elsevier Science Publishers B.V., 1990

All rights reserved. No part of this publication may be reproduced, stored in a retrieval system or transmitted in any form or by any means, electronic, mechanical, photocopying, recording or otherwise, without the prior written permission of the publisher, Elsevier Science Publishers B.V./ Physical Sciences & Engineering Division, P.O. Box 330, 1000 AH Amsterdam, The Netherlands.

Special regulations for readers in the USA — This publication has been registered with the Copyright Clearance Center Inc. (CCC), Salem, Massachusetts. Information can be obtained from the CCC about conditions under which photocopies of parts of this publication may be made in the USA. All other copyright questions, including photocopying outside of the USA, should be referred to the publisher.

No responsibility is assumed by the Publisher for any injury and/or damage to persons or property as a matter of products liability, negligence or otherwise, or from any use or operation of any methods, products, instructions or ideas contained in the material herein.

Printed in The Netherlands

PREFACE

Marine air from off the oceans of the world comes in to the continents bearing slightly increased levels of carbon dioxide and other substances characteristic of a civilized and increasingly industrialized society. The fact that this air contains natural radioactivity in small quantities has been recognized in only relatively recent times.

Present in the air in readily detectable amounts is radon which comes primarily from the soil where very small amounts of its parent, radium, exists. Radon is a naturally radioactive inert gas that becomes a hazard only when found in concentrations such as those encountered in unventilated uranium mines. The fact that it is mildly radioactive, and does not combine with other gases, makes it unique as a tracer for the studying of a number of processes in the indoor and outdoor atmospheres.

The discovery of radon, its characteristics, and sources in the environment as well as methods of control and use in research in the atmosphere are described. Possible health effects from extended exposure at high levels of concentration are treated also.

IN APPRECIATION

The author extends his thanks and heartfelt gratitude to colleagues on the faculty and administration through the years for their interest and appreciation. Financial support for the radon research has come from the Research and Economic Development Division of the New Mexico Institute of Mining and Technology, from the Division of Atmospheric Sciences of the National Science Foundation, and from the Office of Health and Environmental Research of the United States Department of Energy. The help of S.D. Schery in the development of our Radon Laboratory and its programs is gratefully acknowledged.

Sincere appreciation is extended to technical secretary, Pamela Norton, for her skills in word processing, her knowledge of composition, and friendly professional attitude.

In a very special way his warmest thanks and heartfelt gratitude go to his wife, Ruby. Without her help, encouragement and patience, the task would not have been accomplished.

CONTENTS

PREFACE

1. INTRODUCTION...1
   1.1 HISTORICAL PERSPECTIVE......................................1
   1.2 DISCOVERY OF RADON..2
   1.3 THE RADIATION ENVIRONMENT...................................6

2. RADIOACTIVITY...9
   2.1 RADIOACTIVE DECAY...9
       2.1.1 Mean life and half-life..............................10
       2.1.2 Decay series...12
       2.1.3 Relation to nuclear structure........................13
       2.1.4 Radioactive growth and decay.........................17
       2.1.5 Radioactive equilibrium..............................18
       2.1.6 Age determination....................................20

3. CHEMISTRY AND PHYSICS OF RADON.................................25
   3.1 ISOTOPES OF RADON..25
   3.2 PHYSICAL PROPERTIES..26

4. SOURCE OF RADON IN ROCKS AND SOILS.............................29
   4.1 URANIUM, THORIUM AND ACTINIUM..............................29
       4.1.1 Uranium..30
       4.1.2 Thorium..31
       4.1.3 Actinium...32
   4.2 RADIUM - THE PARENT ELEMENT................................32
       4.2.1 Crustal abundance....................................33
       4.2.2 Radium isotopes......................................34
   4.3 EMANATION PROPERTIES.......................................35

	4.4	RADON, THORON AND ACTINON	36
		4.4.1 Radon (222Rn)	37
		4.4.2 Thoron (^{220}Rn)	39
		4.4.3 Actinon (^{219}Rn)	40
5.	RADON - SOIL TO AIR		43
	5.1	TRANSPORT PROCESSES	43
		5.1.1 Diffusion and viscous flow	44
		5.1.2 Flow in channels	48
	5.2	MEASUREMENT	50
		5.2.1 Accumulator	50
		5.2.2 Flow method	52
		5.2.3 Adsorption method	52
		5.2.4 Vertical profile	53
		5.2.5 Soil concentration gradient	54
	5.3	RESULTS OF FLUX DENSITY MEASUREMENTS	55
6.	RADON IN THE ATMOSPHERE		59
	6.1	GROUND LEVEL	59
	6.2	VERTICAL DISTRIBUTION	60
	6.3	CONTINENTAL AND MARINE AIR MASSES	62
	6.4	DIURNAL AND SEASONAL CHANGES	63
		6.4.1 Diurnal variation	63
		6.4.2 Seasonal variation	66
7.	RADON AS A TRACER IN THE ATMOSPHERE		71
	7.1	TRANSPORT PROCESSES	71
		7.1.1 Radon-222	73
		7.1.2 Radon-220 (thoron)	76
8.	RADON DECAY PRODUCTS IN THE ATMOSPHERE		81
	8.1	UNATTACHED RADON DECAY PRODUCTS	82
		8.1.1 Mean life	82
		8.1.2 Mobility of the ^{222}Rn decay product ions	83
		8.1.3 Tracers in the atmospheric electrical environment	84

8.2   DECAY PRODUCTS ATTACHED TO AEROSOLS...................85
        8.2.1   Source and size distributions of aerosols......85
        8.2.2   Number concentration...........................87
        8.2.3   Vertical distribution of aerosols..............88
  8.3   SUMMARY...89

9. EFFECT UPON THE ELECTRICAL CHARACTER OF THE ATMOSPHERE.....91
  9.1   IONIZATION..91
  9.2   CONCENTRATION...92
  9.3   CONDUCTIVITY..94
  9.4   MOBILITY..94
  9.5   SUMMARY...95

10. RADON UNDERGROUND..97
  10.1 RADON IN AN UNDERGROUND CAVITY........................98
  10.2 RADON IN A TUNNEL.....................................99
  10.3 RADON IN MINES.......................................101
        10.3.1 Uranium mining................................102
        10.3.2 Radon and aerosols............................103
        10.3.3 Mines other than uranium......................105

11. RADON IN WATER..109
  11.1 PUBLIC WATER SUPPLIES................................109
  11.2 SEA WATER..110

12. INDOOR RADON..113
  12.1 SOURCES AND TRANSPORT................................113
        12.1.1 Construction materials........................114
        12.1.2 Water supplies................................114
        12.1.3 Natural gas and other sources.................115
  12.2 EQUILIBRIUM AND PLATEOUT.............................116
        12.2.1 Equilibrium fraction..........................116
        12.2.2 Plateout......................................117
  12.3 VENTILATION..117
  12.4 METHODS FOR CONTROL..................................118
        12.4.1 Soil..118

	12.5	SURVEYS..119
		12.5.1 <u>International</u>............................120
		12.5.2 <u>United States</u>............................121
		12.5.3 <u>Radon measurement at the state level</u>.......121
	12.6	SUMMARY..122

13. HEALTH EFFECTS..125
    13.1   RADIATION FROM RADON AND ITS DECAY PRODUCTS.........125
            13.1.1 <u>Quantities and units</u>.......................126
    13.2   THE HUMAN DOSE......................................127
    13.3   IMPACT..128
    13.4   SPAS..129

APPENDIX..131
    A.1   The Uranium Series..................................132
    A.2   The Thorium Series..................................133
    A.3   The Actinium Series.................................134

CHAPTER 1

INTRODUCTION

1.1 HISTORICAL PERSPECTIVE

Radon was first identified as a unique substance in 1900 only four years after the discovery of radioactivity by Henri Becquerel in Paris in 1896. Ernest Rutherford early in 1900 while working with thorium oxide, found that in addition to the ionization from alpha and beta radiations there was an additional ionizing gas diffusing from thicker layers of the thorium compounds. This was the thorium emanation or thoron as it came to be called. A few months later in 1900, Fritz Dorn while studying radium salts observed a similar radioactive gas which he referred to as radium emanation or radon as it is now known. A. Debierne and F. Giesel are credited with the discovery of emanation from actinium in 1903. The actinium was separated from pitchblende by a chemical procedure.

The term "emanation" (symbol Em) was first suggested by Rutherford for these radioactive gases. The word "niton" was used by Ramsay and Gray for the radium emanation because of the luminosity of the new element in the condensed state ("niton" in Latin meaning the "shining one"). In 1923 the International Committee on Chemical Elements approved the words radon, thoron, and actinon for the radioactive gases to show their relation to the parent substances radium, thorium, and actinium.

1.2 THE DISCOVERY OF RADON

The discoveries of the naturally occurring radioactive gases came during a remarkable period in the history of physics which occurred as the 20th century began. Roentgen's work in Germany lead to the discovery in 1895 of penetrating rays that produced luminescence and caused the darkening of photographic plates. He chose to call them X-rays because of their unknown properties, and although they did not play a direct role in the discovery of radioactivity, Roentgen's work did provide an important framework for the discovery of radioactivity which followed within the next few years.

It was the French physicist, Henri Becquerel, who in 1896 performed some ingenious experiments with uranium salts and showed that they emitted rays that persisted over long periods of time and had the unique property of being able to discharge a gold-leaf electroscope. These rays produced ionization and were much more readily absorbed by thin sheets of aluminum than X-rays. It was found in 1899 that unlike the Roentgen rays Becquerel rays from uranium could be deflected by magnetic fields.

The discovery in 1898 of other natural radioactive substances polonium and radium by Pierre and Marie Curie and their assistant G. Bemont in Paris, was made possible by an extraordinary effort on the chemical treatment of pitchblende [uranium oxide and chalcolite (copper, uranium phosphate)]. More than a ton of pitchblende residue was obtained from the Joachimisthal mine in Bohemia (Czechoslovakia) through the cooperation of the Academy of Science in Vienna and the Austrian government. By 1902 Mme. Curie reported having isolated 0.1 g of radium chloride. This work with radium by the Curies represented a supreme effort and in the words of J.J. Thomson in 1903 "indefatigable zeal and experimental skill."

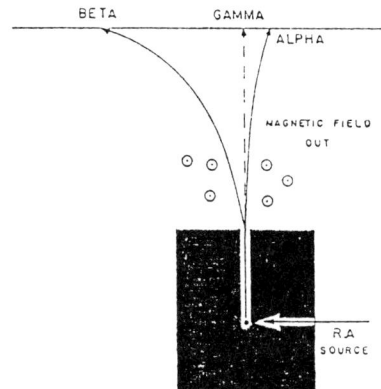

Fig. 1.1 Alpha, beta and gamma rays from a radioactive substance showing their behavior in a magnetic field. From J.M. Cork, Radioactivity and Nuclear Physics, 3rd edn., D. Van Nostrand Co., Princeton, NJ, 1957, p. 14 (ref. 1).

For their work Pierre and Marie Curie and Henri Becquerel received the Nobel Prize in chemistry in 1903. The joy and pride of accomplishment by the Curies and their two little daughters were short-lived when in 1906 Pierre was killed while crossing a busy street in Paris. It was Mme. Curie's careful work with the new element radium that provided Dorn with the material in which radon was discovered. Marie Curie, in 1898, was the first to use the word radioactivity in connection with her study of uranium and thorium compounds in 1898. Marie Curie received a second Nobel prize for her work with radium in 1911.

It was Ernest Rutherford and Frederich Soddy working at McGill University in Canada in the period 1902-1904 who laid the foundations for the theory of the radioactive atom. Rutherford identified and named the alpha and beta rays on the basis of their relative abilities to penetrate aluminum foils. Gamma rays, a very penetrating radiation similar to X-rays, was discovered by P. Villard in France at about this same time. See Fig. 1.1. Later on in 1906 careful measurements of the charge-to-mass ratio of the alpha particle showed that they must either be a singly charged hydrogen molecule or doubly charged helium atoms. Rutherford favored the latter. Working with Soddy he explained natural radioactivity as the spontaneous change of one element to another. For his work he received the Nobel Prize in chemistry in 1908.

An analysis of the energy expended by the alpha and beta particles in ionizing the air led Rutherford to state in his paper on Radioactive Change in 1903, ". . . the energy latent in the atom must be enormous to that rendered free in ordinary chemical change." It was not until 1937 that Hans Bethe worked out the nuclear energy reaction that is the source of the sun's energy, and 1942 when Enrico Fermi produced the first self-sustained nuclear reaction in the Chicago pile which demonstrated the controlled release of energy from the nuclei of atoms.

Rutherford's work in this period was extraordinary not only for laying the foundation for understanding radioactive transformations and identifying the particles involved, but also for working with many talented scientists from neighboring countries. Rutherford was a native of New Zealand. In addition to Soddy in Canada; Becquerel, the Curies, and Villard in France; the list included German scientists: Hans Geiger, Otto Hahn, J. Elster and F. Geitel. The latter were the first to identify radon decay products in ion form in the atmosphere, and F. Giesel studied the effects of magnetic fields on the rays from radium following Rutherford's announcement of the "uranium rays" a few months earlier. Giesel was also involved in the early study of actinium. Geiger along with W. Mueller some 20 years later devised the widely known G-M counter for radiation detection. Otto Hahn with F. Strassman in Berlin in 1939 (31 years later) were co-discoverers of the fission of uranium nuclei.

The eight-year period from 1896 to 1903 becomes especially outstanding when in summary the following discoveries are noted:

1) the inherent instability in the atoms of heavy elements that leads to radioactive decay involving transformations from one element to another,

2) the presence of ionizing radiations, the alpha, beta and gamma rays, having the ability to cause fluorescence in certain minerals and possess widely differing absorption characteristics,

3) the energies in the decay processes exceeding those of chemical binding by factors of a million or more (an increase in temperature due to radioactive decay was actually observed),

4) the identification of the naturally radioactive elements uranium, thorium, actinium, and associated elements in the same minerals along with radium and polonium,

5) the finding of radioactive but chemically inert gases including radon, thoron and actinon, and

6) the detection of helium associated with radioactive minerals.

TABLE 1.1

Important scientific discoveries pertaining to radioactivity and the nature of the atom.

Discovery and Year	The Scientists and Their Locations	Nobel Prize Awarded
X-rays (1895)	W.C. Roentgen (Germany)	1901
Radioactivity (1896)	Henri Becquerel (France)	1903
Electron (1897)	J.J. Thomson (Cavendish Lab., U.K.)	1906
Uranium and Thorium (1898)	Pierre and Marie Curie (Univ. of Paris)	1903
Radium (1900)	Marie Curie (Univ. of Paris)	1911
Radioactive Trans- formations (1902-1904)	E. Rutherford and F. Soddy (McGill Univ. Canada)	1908 1921

These observations had to be taken into account in the final description of the radioactive process. All together there were eight Nobel Prizes awarded to scientists participating in the eight-year period including the discovery of radioactivity. Details are given in Table 1.1.

1.3 THE RADIATION ENVIRONMENT

The first recorded awareness of the effects of environmental radiation on humans was reported by Agricola for miners in the Erz Mountains of eastern Europe in 1556. The first association of lung cancer with miners was determined in 1879. It was not until the 1950's that it was found that radon was the primary cause of lung cancer in miners, however. Lesions or "burns" from X-rays were observed by the Curies within a year following the discovery of these penetrating rays in 1896.

Within a few years of the discovery of radium in pitchblende it was recognized that all soil and rocks emitted alpha, beta, and gamma radiation from uranium, thorium, and other natural radioactive elements in the earth's crust. This led to the recognition of <u>terrestrial radiation</u> as a basic component of our radiation environment. Additional exposure to naturally occurring radionuclides results from the release of ^{40}K, ^{238}U and ^{232}Th from

TABLE 1.2

Estimated effective dose equivalent to the lung, gonads, bones and other tissues for various sources of natural background radiation in the United States and Canada. From NCRP No. 94 (ref. 2).

Source	Total Effective Dose Equivalent (mSv/y)
Cosmic	0.27
Cosmogenic	0.01
Terrestrial	0.28
Inhaled	2.0
Internal	0.40
Total	3.0

coal fired power plants, from phosphate fertilizers, and from building materials. In recent years careful studies have shown increasing amounts of radiation from consumer products.

A significant part of the radiation environment comes from space in the form of cosmic radiation. As early as 1900 many of the early investigators of radioactivity observed a small leakage in their electroscopes. Measurements on mountain tops and in balloons soon pointed the way to the extraterrestrial origin of this interesting radiation. In addition to the effects of the incident cosmic rays themselves, the interaction of these high speed particles with the nuclei of atoms in the atmosphere produces new cosmogenic radionuclides. Chief among these are carbon-14 and tritium(^{3}H). They are relatively long-lived isotopes and are major constituents of body tissue.

A summary of the sources of radiation listed above is shown in Table 1.2. The information in the table clearly shows that inhaled nuclides, primarily radon and its daughters, are by far the major source of radiation to living tissue in the body. Most of this dose including some ^{40}K, ^{87}Pb ^{238}U, and ^{232}Th on dust is delivered to the bronchial epithelium. A further component to human dose comes from food and water that contain natural radioelements. The dose to the body from natural sources will be treated in detail in a later chapter.

Radon and thoron due to their wide distribution in the atmosphere, make up a unique set of tracers for a variety of transport and mixing processes. Measured profiles of these isotopes above the earth's surface are in general agreement with those predicted by turbulent diffusion theory. Nocturnal drainage winds and cumulus convection have been studied using ^{222}Rn and its long-lived daughter, ^{210}P. Turbulent diffusion within the first few meters of the earth's surface have been studied using ^{220}Rn. Its short half-life of 55 seconds is suited to such studies.

Aerosols in the atmosphere furnish another important subject for the application of radon daughter products in the study of size spectra, attachment characteristics, and removal mechanisms.

The ionization by radon and its daughters in the lower atmosphere and its effect upon atmospheric electrical parameters is well known. Knowledge of the mobility and other characteristics of radon daughter ions has led to applications in the study of atmospheric electrical environments under fair weather and thunderstorm conditions. The availability of increasingly sophisticated analytical tools and atmospheric measurement systems can be expected to add much to our understanding of radon and its daughters as trace components of the atmospheric environment.

REFERENCES

FOR FURTHER READING

1   J.M. Cork, Radioactivity and Nuclear Physics, 3rd. edn., D. Van Nostrand Co., Princeton, NJ, 1957, p. 14.
2   NCRP 94, Exposure of the Population of the United States and Canada from Natural Background Radiation, National Council on Radiation Protection and Measurements, Bethesda, MD, 1987, From Table 9.6, p. 148.

CHAPTER 2

RADIOACTIVITY

2.1 RADIOACTIVE DECAY

Early workers dealing with radioactivity observed that the amount of fluorescence, the degree of darkening of photographic plates, or the amount of ionization in air produced by a radioactive substance depended not only upon the quantity present but also upon the distinctive character of a particular element. This led to the basic law of radioactive decay which relates the rate at which atoms disintegrate to the number of atoms present at a given time. The result is

$$\frac{dN}{dt} = -\lambda N \qquad (2.1.1)$$

where N is the number of atoms present at time, t; dN is the number of atoms disintegrating in a time interval dt, and λ is the disintegration constant for a given radioactive element. The minus sign is used to indicate a decrease in N with time. Integration of eqn. (2.1.1) yields

$$\ln \frac{N}{N_o} = -\lambda t \quad \text{or} \quad \frac{N}{N_o} = e^{-\lambda t} \quad (2.1.2)$$

where N_o is the number of atoms present at an arbitrary initial time, $t = o$ and N is the number present at time, t. Equation (2.1.2) shows that N decreases exponentially with time and also that "daughter" atoms resulting from the decay of the original radioactive material, (N_o-N), will increase at an exponential rate $[N_o(1-e^{-\lambda t})]$. As a matter of interest the escutcheon presented to Rutherford when he became Lord Rutherford contained a graph showing these relations (ref. 1).

In the laboratory it is the activity, $A = dN/dt$, that is determined by counting the alpha and/or beta and gamma rays from a given sample by use of appropriate detectors and electronic counting equipment. Hence, it is convenient to use equation (2.1.2) in the following forms:

$$\ln \frac{A}{A_o} = -\lambda t \quad \text{and} \quad A = A_o e^{-\lambda t} \quad (2.1.3)$$

The number of atoms N in a pure sample of known mass of a given radioactive element can be determined if necessary from Avogadro's number and the mass number. In practice one usually deals with the activity, A.

2.1.1 Mean life and half-life

The mean life, τ, of a radioactive element is the average time that a radioactive atom can be expected to exist before it decays.

From equation (2.1.1)

$$\lambda = -\frac{dN/N}{dt} \qquad (2.1.4)$$

which shows that the decay constant is the fraction of atoms decaying per unit time, or the probability of decay per unit time since this is a statistical process. The reciprocal is then the average life of the atom, i.e., $\tau = 1/\lambda$. In equation (2.1.3) τ is the value of t required for the number of atoms N in a sample to reduce by a factor of 1/e whence

$$\frac{N}{N_o} = \frac{1}{e} = e^{-\lambda \tau} \qquad (2.1.5)$$

consistent with $\tau = 1/\lambda$ as stated above.

The half-life T, of a radioactive substance is defined as the time required for an original number of atoms N_o to reduce to one-half that number, i.e., $N = N_o/2$. Substituting in equation (2.1.2) yields $\ln 2 = \lambda T$ and

$$T = \frac{0.693}{\lambda}. \qquad (2.1.6)$$

If common logarithms (base 10) are used in equation (2.1.3) it can be expressed as:

$$\log A = \log A_o - 0.434\lambda t \qquad (2.1.7)$$

If then the activity of a sample is plotted as a function of time on semilog paper, a straight line results and λ can be calculated from the above. The decay constant (or half-life) of short-lived elements is determined directly from equation (2.1.1) by careful measurements of the mass (no. of atoms) and the activity dN/dt, of a small sample.

The half-lives of natural radioactive elements vary from 1.41×10^{10} y for thorium-232 to 2.92×10^{-7} s for polonium-212.

2.1.2 Decay series

The Curies in 1899 while working with radium observed what they called "excited radioactivity" or an "active deposit" produced on neighboring bodies. Rutherford the following year in his studies with thorium found the same phenomena associated with thorium emanation (thoron). Similar active deposits were found associated with actinium emanation discovered by A. Debierne in 1903. The activities from these deposits changed with time and analyses proved that the elements involved were indeed different from the parent substances.

These early observations led eventually to a listing of the uranium, thorium, and actinium decay series. The main characteristics of these series are given in Table 2.1. Each series begins with a natural radioelement of very long half-life of the order of the age of the earth at about 4.5 billion (4.5×10^9) years. Since the mass lost in the decay processes is due almost exclusively to emission of alpha particles (helium nuclei) with a mass of 4, the mass numbers of the members of a given radioactive series can be represented by 4 times an integer, n, plus an appropriate number less than 4. For example, the uranium series is the 4n + 2 group: for ^{238}U, n = 59; for ^{222}Rn, n = 55, etc. Thorium is the 4n series; and actinium, the 4n + 3 series. Although the 4n + 1 series is not found in nature in identifiable amounts, it has been produced in the laboratory with neptunium-237 as its long-lived (2.2×10^6 y) element. If present in the earth's crust at the time of formation, it would have decayed by now.

Another characteristic common to all of the natural series is the existence of an isotope of radon, element 86. In the uranium series there is ^{222}Rn (radon); in the thorium series ^{220}Rn (thoron);

TABLE 2.1
Some characteristics of the natural radioactive decay series.

Series	Uranium	Thorium	Actinium
Mass number code	4n + 2	4n	4n + 3
Long-lived parent and half-life	^{238}U 4.51 X 10^9 y	^{232}Th 1.39 X 10^{10} y	^{235}U 7.13 X 10^8 y
Radium parent and half-life	^{226}Ra 1,600 y	^{223}Ra 11.4 d	^{224}Ra 3.66 d
Radon isotope and half-life	^{222}Rn (radon) 3.82 d	^{220}Rn (thoron) 55.6 s	^{219}Rn (actinon) 4.0 s
Potential alpha energy in short-lived radon decay chain* (OECD-1983)	19.2 Mev per atom	20.9 Mev per atom	20.8 Mev per atom
Stable end product	^{206}Pb	^{208}Pb	^{207}Pb

*The potential alpha energy is the total alpha energy emitted during the decay of an atom along its decay chain.

and in the actinium series ^{219}Rn (actinon). There is no isotope of radon in the artificially produced neptunium series. Yet another characteristic of the natural series is that they all end with a stable isotope of lead: ^{206}Pb, ^{207}Pb, and ^{208}Pb for the uranium, actinium, and thorium series respectively.

A detailed listing of the natural radioactive decay series is given in Appendix A.1, A.2, and A.3.

2.1.3 Relation to nuclear structure

The periodic table of the elements (Fig. 2.1) lists all atomic species from hydrogen (1) through uranium (92) and beyond according to the number of electrons and their arrangement in shells surrounding the nucleus. The deficiency or surplus of electrons

THE PERIODIC TABLE

	I		II		III		IV		V		VI		VII		0 or VIII	
	a	b	a	b	a	b	a	b	a	b	a	b	a	b	a	b
1	1 H 1.0078															2 He 4.002
2	3 Li 6.940		4 Be 9.02		5 B 10.82		6 C 12.01		7 N 14.008		8 O 16.0000		9 F 19.00		10 Ne 20.183	
3	11 Na 22.997		12 Mg 24.32		13 Al 26.97		14 Si 28.06		15 P 31.02		16 S 32.06		17 Cl 35.457		18 A 39.944	
4	19 K 39.096		20 Ca 40.08		21 Sc 45.10		22 Ti 47.90		23 V 50.95		24 Cr 52.01		25 Mn 54.93		26 Fe 55.84	27 Co 58.94	28 Ni 58.69
	29 Cu 63.57		30 Zn 65.38		31 Ga 69.72		32 Ge 72.60		33 As 74.91		34 Se 78.96		35 Br 79.916		36 Kr 83.7	
5	37 Rb 85.48		38 Sr 87.63		39 Y 88.92		40 Zr 91.22		41 Cb 92.91		42 Mo 96.0		43 —		44 Ru 101.7	45 Rh 102.91	46 Pd 106.7
	47 Ag 107.880		48 Cd 112.41		49 In 114.76		50 Sn 118.70		51 Sb 121.76		52 Te 127.61		53 I 126.92		54 Xe 131.3	
6	55 Cs 132.91		56 Ba 137.36		57-71 La 138.92 Rare Earths		72 Hf 178.6		73 Ta 180.88		74 W 184.0		75 Re 186.31		76 Os 191.5	77 Ir 193.1	78 Pt 195.23
	79 Au 197.2		80 Hg 200.61		81 Tl 204.39		82 Pb 207.21		83 Bi 209.00		84 Po 210		85 —		86 Rn 222	
7	87 —		88 Ra 226.05		89 Ac 227		90 Th 232.12		91 Pa 231		92 U 238.07					

Fig. 2.1 The periodic table of the elements. The atomic number is just above the symbol for each element, and the atomic weight averaged over all natural isotopes for the element is below the symbol.

in the shells beyond those needed to match the positively charged protons in the nucleus determine the chemical affinity of one atom for another. Atoms with filled shells such as helium, argon, and radon are essentially chemically inert. The classification of the elements in the periodic chart have been very successful as a guide in understanding the chemical properties of the elements. We now proceed to a description of the atomic nucleus and its transformations as found in the natural radioactive decay series.

The <u>atomic number</u> is determined by the number of protons present in the nucleus, e.g. one for hydrogen, 12 for carbon, 86 for radon, 92 for uranium, etc. Each proton carries a unit positive charge exactly balancing the negative charge in each electron in the external electronic structure. While the charges are equal and opposite in sign, the proton is about 1,835 times as massive as the electron. The other major constituent of the nucleus is the neutron which carries no charge but has a mass the same as that of the proton within a tenth of 1%. Hence, element X can be designated by $^A_Z X$ where Z is the <u>atomic number</u> or number of protons and A = Z + N where A is the mass number and N is the number of neutrons in the nucleus. For example, radon-222 is $^{222}_{86}Rn$ which means that there are 86 protons and (222-86) 136 neutrons in the nucleus.

The alpha decay of element X can be represented by $^A_Z X - ^4_2 He = ^{A-4}_{Z-2} Y$ since the alpha particle is a helium nucleus. The decay of $^{222}_{86}Rn$, an alpha emitter yields $^{218}_{84}Y$ and since the periodic table (Fig. 2.1) shows that element 84 is polonium, we have $^{222}_{86}Rn = ^{218}_{84}Po + ^4_2 He + Q$ where Q is the energy released resulting from the fact that the combined masses of the polonium and helium atoms are slightly less (Δm) than that of the Rn atom. Q then is equal to Δmc^2 from Einstein's mass-energy relationship.

Beta decay processes are ones in which a neutron in the nucleus spontaneously converts to a proton with the emission of an electron. Hence, the decay of element X by beta emission can be expressed as $^A_Z X = ^A_{Z+1} Y + ^{-1}_0 e + \nu$ where the mass number A stays the same and the conversion of a neutron to a proton requires the addition of one positive charge which is provided by the emission of the electron carrying a minus charge. The neutrino, ν was proposed by W. Pauli and E. Fermi in 1934 in the theory of beta decay in order to conserve energy and momentum in the process. Again a very minute mass difference yields mc^2 energy which is available as kinetic energy for the electron and neutrino. In the ^{222}Rn decay chain $^{214}_{82}Pb$ undergoes beta decay to yield $^{214}_{83}Bi$ plus a neutrino and an electron having an energy of 0.65 X 10^6 electron-volts, enough for it to penetrate about 10 m of air at standard pressure before coming to rest.

The nuclei of atoms are extremely small with a radius of the order of 10^{-14} m compared with the atom as a whole which is about 10,000 times larger at approximately 2×10^{-10} m. In order for the nuclei to retain their stability the protons and neutrons must have a unique binding force much greater even than that required to overcome Coulomb repulsion among the closely packed protons. A major undertaking of physics in recent decades has been the search for an understanding of nuclear forces. A major advance was made in 1935 when H. Yukawa proposed the <u>meson</u> with a mass approximately 200 times the electron mass as the "field" particle for the strong forces within the nucleus. Structure and binding energy has resulted in the identifying of four force fields: the <u>strong</u> interaction between nucleons (protons and neutrons), the <u>electromagnetic</u> or Coulomb-type interaction, the <u>weak</u> interactions including the electron and neutrino in beta decay, and <u>gravity</u>. The latter is inconsequential on the atomic/nuclear scale. Altogether more than 200 elementary particles including antiparticles and "resonances" are known. The advent of careful studies of cosmic rays and the building of high energy accelerators from the Van de Graaf generator and cyclotron to the Super Conducting Super Collider are the instruments used to accelerate particles to very high energies which are needed for understanding of the basic nuclear forces.

From an examination of the nature of nuclear structure it becomes apparent that as one goes through the periodic table very few naturally occurring radioactive elements are encountered until the lead-uranium section is reached. Thus, one can infer that an inherent instability occurs in the range of atomic (proton) numbers of 82-92 and mass numbers (neutrons plus protons) 204 to 238. Natural radioactivity, especially as it is exhibited in the uranium, thorium, and actinium series, is a progression from less stable to more stable nuclei.

2.1.4 Radioactive growth and decay

When one or more of the decay products of a radioelement are themselves radioactive a radioactive series results which with certain simplifying assumptions can be treated as follows. Let N be the number of atoms of the original parent present at time t and the initial number N_{10}. Then let the number of atoms of its decay product be N_2 at time t with $N_2 = 0$ at $t = 0$. Then the number of daughter atoms being formed in time dt is just the number of the parent atoms undergoing disintegration in the same interval, and the number of daughter atoms decaying is $\lambda_2 N_2$. Therefore, the net change in daughter atoms is

$$\frac{dN_2}{dt} = \lambda_1 N_1 - \lambda_2 N_2 \tag{2.1.8}$$

From equation (2.1.2) $N_1 = N_{10} e^{-\lambda_1 t}$. Substituting in (2.1.8) and multiplying by $e^{\lambda_2 t} dt$ gives

$$e^{\lambda_2 t} dN_2 + \lambda_2 N_2 e^{\lambda_2 t} dt = \lambda_1 N_o e^{(\lambda_2 - \lambda_1)t} dt \tag{2.1.9}$$

Integrating yields

$$N_2 e^{\lambda_2 t} = \frac{\lambda_1}{\lambda_2 - \lambda_1} N_{o1} e^{(\lambda_2 - \lambda_1)t} + C \tag{2.1.10}$$

If $N_2 = 0$ at $t = 0$

$$0 = \frac{\lambda_1 N_{o1}}{\lambda_2 - \lambda_1} + C \tag{2.1.11}$$

Solving (2.1.11) for C and substituting in (2.1.10) gives

$$N_2 e^{\lambda_2 t} = \frac{N_{o1} \lambda_1}{\lambda_2 - \lambda_1} (e^{(\lambda_2 - \lambda_1)t} - 1) \tag{2.1.12}$$

Solving (2.1.12) for N_2, the number of daughter atoms as a function of time t gives

$$N_2 = \frac{N_{o1}\lambda_1}{\lambda_2 - \lambda_1}(e^{-\lambda_1 t} - e^{-\lambda_2 t}) \qquad (2.1.13)$$

This relation for a parent having a half-life of 1 hour and a daughter of half-life 5 hours is shown in Fig. 2.2. It will be observed also that N_2 is zero (as assumed) at t = o and also at t = ∞ as suggested by curve B in the figure.

Equation (2.1.13) is only the first step in a radioactive decay series. The use of as many differential equations or an appropriate matrix as there are elements gives the Bateman equations which can be used to find the number or activity of any member of the series at a time t provided that the initial number of atoms in each case except the first was zero.

2.1.5 Radioactive equilibrium

Secular. When the wide range in half-lives of the elements in the natural radioactive decay series is considered, calculating the number of atoms for a given element in the series can be simplified in certain cases. For example ^{226}Ra has a half-life of 1,620 y while its daughter ^{222}Rn has a half-life of only 3.8 d. In this example $T_1 \gg T_2$ and it follows from $T = \frac{\ln 2}{\lambda}$ (2.1.6) that $\lambda_1 \ll \lambda_2$. After a time t that is much longer than T_2 the half-life of the daughter, $e^{-\lambda_2 t}$ approaches zero and $\lambda_2 - \lambda_1 \cong \lambda_2$. Using this in 2.1.13 and remembering that $N_1 = N_{o1} e^{-\lambda_1 t}$ the result is:

$$\lambda_1 N_1 = \lambda_2 N_2 \qquad (2.1.14)$$

Hence, when a long-lived parent and a short-lived daughter are present together for a sufficient length of time, the decay rate of the daughter equals the decay rate of the parent. In the example cited above if ^{226}Ra is in a sealed container with ^{222}Rn, an

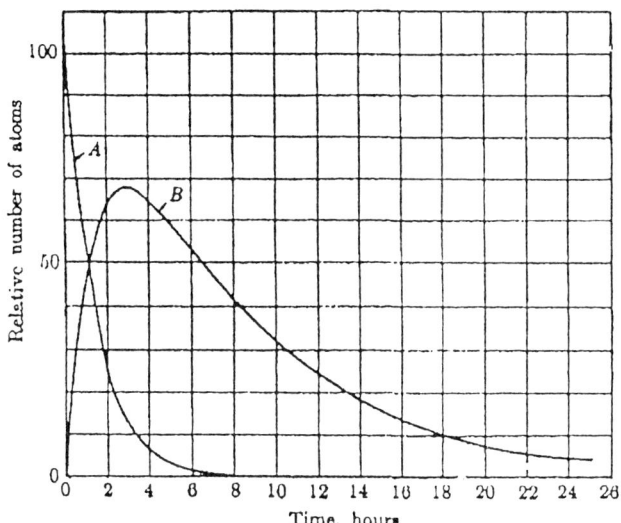

Fig. 2.2 Decay of element A (T = 1 hr) and the growth and decay of element B (T = 5 hr). From Modern University Physics, Addison-Wesley, Reading, MA, 1960, p. 875 (ref. 4).

equilibrium state will be attained where the activities are identical. This phenomenon is called <u>secular equilibrium</u>. If there are a number of radioactive elements in succession the general condition for secular equilibrium is given by

$$\lambda_1 N_1 = \lambda_2 N_2 = \lambda_3 N_3 = \text{----} \lambda_n N_n \qquad (2.1.15)$$

<u>Transient</u>. If the parent radioactive element has a half-life that is longer than the daughter but only moderately so (x 10), then $T_1 > T_2$ ($\lambda_1 < \lambda_2$) and equation (2.1.13) can be written as

$$\lambda_1 N_1 = (\lambda_2 - \lambda_1) N_2 \qquad (2.1.16)$$

where again, N_{o2} is zero at $t = o$ and $t \gg T_2$. These conditions apply to what is called <u>transient</u> equilibrium. Just as in the case of secular equilibrium both parent and daughter activities decrease at equal rates, but the rate of decline of activity depends upon the T of the parent. However, since the half-life of the parent is much less than in the case of secular equilibrium, there is a pronounced decrease in total activity with time. An example of transient equilibrium occurs in the case of ^{222}Rn and its daughters (the active deposit).

No equilibrium. This condition exists when the half-life of the parent is shorter than that of the daughter. Starting with a pure parent, the activity of the daughter increases, passes through a maximum and then decreases which follows from equation (2.1.13) with $N_{o2} = 0$. The time of maximum activity (t_m), can be obtained from the requirement that $dN_2/dt = o$ in equation (2.1.13) where as in transient equilibrium $t \gg T_2$. The result is

$$t_m = \frac{\ln \lambda_2 - \ln \lambda_1}{\lambda_2 - \lambda_1} \qquad (2.1.17)$$

For example there could be no equilibrium between ^{218}Po (3.05 min), the first daughter of ^{222}Rn, and ^{214}Pb (26.8 m) the second daughter. From an initially pure sample of ^{218}Po the maximum activity (t_m) would occur at 10.8 m following $t = o$.

Examples of secular, transient, and no equilibrium cases are given in Fig. 2.3.

2.1.6 Age determination

A notable application of the theory of radioactive decay series is the determination of the age of rocks and minerals containing uranium (^{236}U), thorium (^{232}Th), and actinium (^{235}U). As noted in Table 2.1 these series have a long-lived parent and end up with a stable isotope of lead. Uranium-238 for example is found to the extent of a few parts per million in the rocks and soil of the earth's crust. A given specimen of rock should contain atoms of ^{206}Pb in an amount equal to all the ^{238}U atoms which have decayed since that rock was formed.

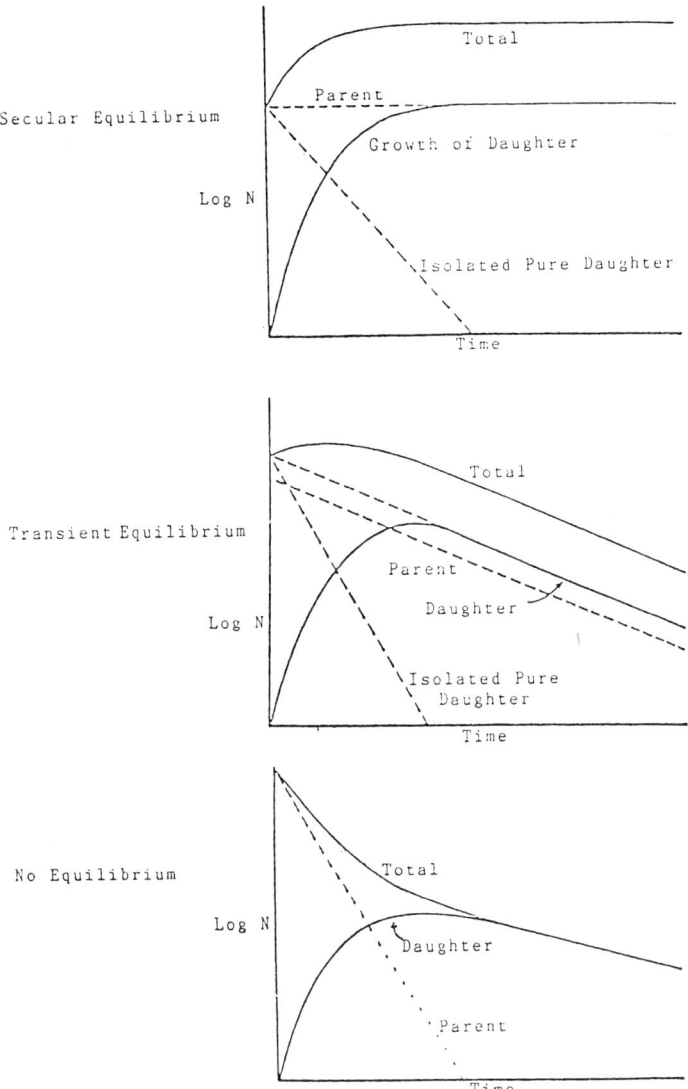

Fig. 2.3 Secular and transient equilibria and an example of no equilibrium. The logarithm of the number of atoms of the parent and daughter elements is given on the ordinate. Adapted from G.D. Chase, Radioisotope Methodology, Burgess Publishing Co., Minneapolis, MN, 1959, pp. 42-44 (ref. 5).

Since equation (2.1.13) gives the number of daughter atoms as a function of time t, it can equally well be applied to any member of a decay series that is in secular equilibrium. Hence, we can let N_{Pb} represent the number of lead atoms present in the sample and $\lambda_2 = \lambda_{Pb} = 0$ (since $T_{Pb} = \infty$ for a stable isotope). Also one can use $N_u = N_{ou} e^{-\lambda_u t}$ (where $N_u = N_1$, etc.) as the number of ^{238}U atoms in the sample, then 2.1.13 becomes

$$N_{Pb} = N_{ou}(1 - e^{-\lambda_u t}) \qquad (2.1.18)$$

If we assume as above that the only reason that lead is present is that it came from the original uranium, then

$$N_{ou} = N_u + N_{Pb} \qquad (2.1.19)$$

Substituting this in equation (2.1.18) and solving for t gives

$$t = \frac{1}{\lambda_u} \ln \left(\frac{N_{Pb} + N_u}{N_u} \right) \qquad (2.1.20)$$

The age of a rock can be calculated from this equation based upon the amount of ^{226}Pb and ^{238}U present in a given sample at the time t. Similarly the appropriate isotope, from the thorium and actinium series can be used in age determinations.

The result of the analyses of many rock samples from widely separated regions of the earth have resulted in a mean value of about 4.5 billion years in general agreement with age determination by other means including stratigraphic and paleontologic methods.

Caution must be exercised in the use of equation (2.1.20) due to certain assumptions involved. For example, 1) What is the evidence for the rock having existed in an undisturbed state over the period in question? 2) Were any contaminating amounts of the stable daughter present originally or might they have migrated through the rock matrix? 3) Is there a chance that the isotopes of the inert gas, radon, present in those series might have escaped? 4) Have the spectrochemical (or other) analyses of the samples been able to rule out possible exchange of intermediate members of the separate decay series in the same rock sample? These and other questions may severely restrict the validity of radioactive methods of geochronology. A detailed analysis of age determination of materials in the earth's crust is given in Physics and Geology, by Jacobs, et al. (ref. 6).

REFERENCES

1   S. Glasstone, Sourcebook on Atomic Energy, 3rd edn., D. Van Nostrand, Princeton, 1967, 140.
2   E. Browne and R.B. Firestone, Table of Radioactive Isotopes (V.S. Shirley, Ed.) Wiley Interscience, New York, NY, 1986.
3   NCRP No. 97, Measurement of Radon and Radon Daughters in Air, National Council on Radiation Protection and Measurements, Bethesda, MD, 1988, 12.
4   Modern University Physics, Fields, Waves, and Particles, Addison-Wesley Publishing Co., Reading, MA, 1960.
5   G.D. Chase, Radioisotope Methodology, Burgess Publishing Co., Minneapolis, MN, 1959, 35-45.
6   J.A. Jacobs, R.D. Russell and J. Tuzio Wilson, Physics and Geology, McGraw-Hill Inc., New York, 1959, pp. 168-203.

CHAPTER 3

CHEMISTRY AND PHYSICS OF RADON

The prominent role of radon among all natural radioelements is due to the fact that it basically is an inert gas. Once formed from one of the natural radioactive series in the Earth's crust it is free to diffuse into soil air and then to the atmosphere by pressure driven flow or further diffusion.

3.1 ISOTOPES OF RADON

Radon is element 86 in the periodic table.

As shown in Table 2.1 ^{222}Rn originates in the ^{238}U decay series and has a half-life of 3.82 days, ^{220}Rn (thoron) is in the ^{232}Th chain with a half-life of 55.6 sec and, ^{219}Rn (actinon) is in the ^{235}U series with its half-life of 4.0s. All are alpha particle emitters. Actinon, ^{219}Rn, due to its short half-life and the comparative scarcity of its long-lived parent, ^{235}U, can usually be omitted from any radiological considerations.

Beyond these isotopes of radon in the decay series a total of 23 other isotopes have been identified ranging from ^{199}Rn to ^{226}Rn. The longest lived of these is ^{211}Rn (15 hr) and most decay via alpha emission but beta emission and orbital electron capture are included (ref. 1).

3.2 PHYSICAL PROPERTIES

Radon is a colorless gas with a density of 9.73 g/l under standard conditions making it the heaviest gas in nature. When cooled below its freezing point, radon has a brilliant phosphorescence which becomes yellow at lower temperatures and orange-red at the temperature of liquid air (ref. 2). It was this property that led radon to be called niton (the shining one) at the time of its discovery.

The radon atom possesses a stable closed shell electronic configuration which gives it the chemical properties of a noble-gas element. It behaves as expected by comparison with the other inert gases in the periodic table including helium, neon, argon, krypton, and xenon. The electronic configuration of neutral radon atoms in the ground state is $5s^2\ 5p^6\ 5d^{10}\ 6s^2\ 6p^6$ ($^1S0$) (ref. 3). The spectrum of radon resembles that of the other closed shell elements.

The electronic structure of radon suggests very limited chemical activity; however, the relatively low first-ionization potential of 10.7 ev suggests some interactions might be possible. It has been reported that fluorine reacts with radon to produce radon fluoride and that clathrates are formed where radon atoms are found within crystal lattices of certain hydrogen compounds (ref. 4). Radon is sometimes referred to as a "metalloid" an element which lies on a diagonal between the true metals and nonmetals in the periodic table. It has some of the characteristics of both groups behaving similarly to boron, germanium, antimony and polonium in this respect (ref. 5).

Some physical properties of radon are given in Table 2.1. Radon is readily absorbed on charcoal, silica gel and similar substances, - a property which can be used to separate it from other gases. Radon can be effectively removed from a sample air stream by collecting it on activated charcoal cooled to the temperature of solid CO_2 (-78.5°C) (ref. 6). Radon is desorbed from charcoal by heating to 350°C.

The relatively high solubility of radon in water (230 cm^3 kg^{-1} at 20°C) accounts for its presence at substantial amounts in certain spring waters. A detailed list of the physical properties of radon is given in Table 3.1 (ref. 7).

TABLE 3.1
Physical properties of ^{222}Rn. Condensed from NCRP Report No. 97 (1988).

Density at 0°C and 1 atm	9.73 g l^{-1}
Boiling point, normal (1 atm)	-62°C
Density of liquid at normal boiling point	4.4g cm^{-3}
Diffusion coefficient in free air	0.1 cm^2 sec^{-1}
Viscosity at 1 atm pressure and 20°C	229.0 micropoise
Critical pressure	62 atm
Critical temperatures	105°C
Solubility in water at 1 atm partial pressure and 20°C	230 cm^3 (STP) kg^{-1} water
Solubility in various liquids at 1 atm pressure and 18°C	
glycerine	0.21 cm^3 kg^{-1} liquid
ethyl alcohol	7.4 cm^3 kg^{-1} liquid
petroleum (liquid paraffin)	9.2 cm^3 kg^{-1} liquid
toluene	13.2 cm^3 kg^{-1} liquid
carbon disulfide	23.1 cm^3 kg^{-1} liquid
olive oil	29.0 cm^3 kg^{-1} liquid

FOR FURTHER READING

1   F. Weigel, Chemiker Zeitung, 102, 1978, pp. 287-299.
2   CRC Handbook of Chemistry and Physics, 60th edn., 1979-80, B-19.
3   A.C. Wahl and N.A. Bonner, Radioactivity Applied to Chemistry, John Wiley and Sons, New York, NY, 1951, pp. 204-207.
4   J.C. Bailor, H.J. Emeléus, R. Nyholm and A.F. Trotman-Dickenson, Comprehensive Inorganic Chemistry, Pergamon Press, New York, NY, 1973, pp. 328-330.

5   L. Stein, Chemical properties of radon, in: P. Hopke (Ed.) Radon and Its Decay Products, American Chemical Society, Washington, DC, 1987, pp. 240-251.
6   H.F Lucas, Improved low-level alpha-scintillation counter for radon, Rev. Sci. Instrum., 28 (1957), 680-683.
7   NCRP Report No. 97, Measurement of Radon and Radon Daughters in Air, National Council on Radiation Protection and Measurements, Bethesda, MD, 1988, 15.

CHAPTER 4

SOURCE OF RADON IN ROCKS AND SOILS

The origin of radon and thoron in the Earth's crust stems directly from the uranium and thorium and their decay products distributed in minute amounts in the ground within a few meters of the Earth's surface. Once formed in or on the rocks and soil particles the radon atoms must reach the air in the soil capillaries before they can be transported to the atmosphere through diffusion or pressure-induced flow. These transport processes will be treated in the next chapter. Attention is given now to the parent elements in the radon, thoron, and actinon decay series and to their emanation properties.

4.1 URANIUM, THORIUM AND ACTINIUM

The amounts of radon, thoron and actinon in the atmosphere depend primarily upon the concentration of uranium and thorium in the soil and rocks. The isotopes involved are the long-lived nuclides at the head of each of the natural radioactive decay series: ^{238}U for the uranium series, ^{232}Th for the thorium series and ^{235}U for the actinium series (Table 2.1). The relative abundances of the isotopes in natural uranium by weight are ^{238}U - 99.28%, ^{235}U - 0.71% and ^{234}U - 0.0054% (ref. 1). Hence, the uranium series headed by ^{238}U can be expected to predominate over the actinium series (^{235}U) in the environment. The ^{234}U is an intermediate member of the ^{238}U series. On the other hand, ^{232}Th, head of the thorium series is even more abundant than ^{238}U in the earth's crust. It is of interest to note that the half-life of ^{238}U is 4.51 x 10^9 y which is nearly equal to the estimated age of the earth. Hence, about half of the uranium present originally has now decayed.

4.1.1 <u>Uranium</u>

Uranium is widely distributed throughout the earth's crust to the extent of about three parts per million (See Table 4.1). Minerals of commercial value containing several thousand parts per million (ppm) of uranium are ordinarily oxides such as uraninite and carnotite with phosphates and monazite sands being of importance in some cases. Uranium ores of commercial grade are found in the Belgian Congo, Czechoslovakia, Canada, the Colorado plateau of the western United States including the Grants Mineral Belt in New Mexico and elsewhere around the world. Rich ores will contain from 1% to 4% uranium while the majority mined in the United States have averaged a few tenths of 1% uranium.

It should be noted that phosphate rocks widely used as agricultural fertilizers contain quantities of uranium of from 8 to 400 ppm with a mean value of about 40 ppm. In general ^{226}Ra is in equilibrium with the ^{238}U; however, an appreciable fraction of the ^{226}Ra is removed as a byproduct in preparation of the fertilizer. The dose received by the general population from this source is not considered significant (ref. 2).

Uranium-234 and ^{230}Th in the uranium series have half-lives of 2.5 X 10^5 y and 8 X 10^4 y respectively (Table 4.1) - ample time for chemical and geologic processes to affect their concentration. On the other hand the intermediate elements in the thorium series have relatively short half-lives; the longest is ^{228}Ra at 6.7 y. Natural physical and chemical processes in the crustal environment have little influence on the concentrations in this case.

TABLE 4.1

Concentrations of uranium-238 and thorium-232 in rocks and soils. From NCRP Report No. 94 (1987) (ref. 3).

Rock Type	^{238}U		^{232}Th	
	ppm	Bq/kg	ppm	Bq/kg
Igneous				
Basalt	0.5-1	7-10	3-4	10-15
Granite	3	40	17	70
Sedimentary				
Shale, Sandstone	3.7	40	12	50
Carbonate rocks	2	25	2	8
Continental upper crust				
(average)	2.8	36	10.7	44
Soils	1.8	66	9	37

4.1.2 Thorium

The head of the thorium series is ^{232}Th which has a half-life of 14.1×10^9 y. While there are other thorium isotopes present in the decay chains of ^{232}Th and ^{238}U, they make up only very small fractions of the total thorium present in rocks and soil.

Thorium ranks just under lead in abundance in the crust at 11 ppm. It is obtained chiefly from monazite sands in India, Brazil, the Soviet Union and from areas in North Carolina and Virginia in the United States. Monazite sands contain approximately 10% thorium providing high external dose rates for residents of coastal areas where they are found. It is used in magnesium-thorium alloys in aircraft and space vehicles in addition to gas light mantles and for certain phosphors in welding (ref. 4).

A summary of concentrations of the primary thorium isotope (^{232}Th) is given in Table 4.1. Thorium-232 exceeds ^{238}U by a factor of from 3 to 5 in many rock types. Granite rocks exceed basalts in the igneous category and shales are more productive than carbonates (e.g. limestones) in the sedimentary class. As noted in the table the crustal abundance of ^{238}U is estimated to be 2.8 ppm (36 Bq kg^{-1}) while ^{232}Th is present in soil and rocks at about 10.7 ppm (44 Bq kg^{-1}). Hence, although ^{232}Th exceeds ^{238}U in the continental crust by a factor of 3.8, the activities are almost equal. Thorium-232 activity exceeds that of ^{238}U by only about 20%.

4.1.3 Actinium

The long-lived head of the actinium series is ^{235}U (7.1 X 10^8 y). This nuclide is sometimes referred to as actinouranium. As mentioned earlier ^{235}U makes up only 0.71% of natural uranium. This plus the fact that the radon isotope in the series, ^{219}Rn, has a half-life of only 3.92 s makes the actinium series of little concern compared with the ^{238}U and ^{232}Th series. Of the small quantity of actinon produced in soil and rocks only a small fraction can diffuse into the soil capillaries and then be transported to the air-earth interface with such a short half-life (3.92 s).

Some readers may recognize ^{235}U as the fissionable isotope used in the nuclear weapon dropped on Hiroshima near the end of World War II. Kilogram quantities of ^{235}U were separated in a large "diffusion" plant in Oak Ridge, Tennessee. Now the most widely used substance in nuclear reactors and weapons is plutonium-239 (^{239}Pu). This element has atomic number 94 and is produced from the bombardment of ^{238}U with neutrons which yields the intermediate nuclide, neptunium-239. Neptunium (^{239}Np) is radioactive, decaying with a half-life of 2.33 d to ^{239}Pu with the emission of a beta particle (electron). Since ^{238}U is approximately 140 times as abundant in nature as ^{235}U, the production of fissionable material in the form of ^{239}Pu from ^{238}U is much more feasible than the separation of ^{235}U.

4.2 RADIUM - THE PARENT ELEMENT

The most direct clue to the production of radon, thoron and actinon is the concentration of the immediate parent isotopes -

^{226}Ra (1,620 y) for radon, ^{224}Ra (3.64 d) for thoron, and ^{223}Ra (11.7 days) for actinon. The half-lives are given in parentheses. The ^{226}Ra content of a soil or rock sample can be determined by gamma ray spectrometry or by an analysis of the amount of ^{222}Rn in equilibrium with the ^{226}Ra in a liquified sample. The ^{224}Ra can be determined with negligible error by assuming that it is in secular equilibrium with the long-lived parent ^{232}Th, or by alpha spectrometry. The latter method may be used for determination of the ^{223}Ra as well.

The ^{226}Ra, ^{224}Ra and ^{223}Ra distributed in crustal soil and rocks release isotopes of the inert gas radon to capillaries and pores where it can reach levels of the order of 1,000 times that of ordinary outdoor air. The result is a steady transport of radon from soil to air.

4.2.1 Crustal abundance

Crustal rocks and soils contain about 40 Bq kg^{-1} of ^{226}Ra with granites generally exceeding limestone and sandstones by a factor of three or so. Building materials reflect concentrations of their crustal sources with concrete ranging up to 70 Bq kg^{-1} and bricks even higher. See (ref. 5).

The results of measurements of ^{226}Ra concentrations in surface soils by the New Mexico Tech laboratory are given in Table 5.1 along with values for ^{222}Rn flux densities at the various locations. The mean value of 38 ± 14 Bq kg^{-1} for ^{226}Ra concentration in soils is in agreement with the NCRP 97 figure of 40 Bq kg^{-1}. A study currently under way in seven provinces in China yields a mean concentration of ^{226}Ra in soil of 39.0 Bq kg^{-1} and a range of 8.2 to 43.8 Bq kg^{-1} in a program involving more than 1,500 samples (ref. 6).

A study of radionuclides in surface soils by Myrick et al. (ref. 7) covering 356 sites in 33 states in the United States gave concentrations of 41 ± 18 Bq kg^{-1} for ^{226}Ra, 37 ± 31 Bq kg^{-1} for ^{238}U and 36 ± 17 Bq kg^{-1} for ^{232}Th. Errors indicated are 2 σ values from the arithmetic mean. Radioactive equilibrium within the uranium decay series is indicated by a close correlation between ^{238}U and ^{226}Ra for the country as a whole. The result for ^{232}Th of 36 Bq kg^{-1} is in good agreement with 37 Bq kg^{-1} given in Table 4.1 for soils.

The chemical behavior of radium in soils is sufficiently different from uranium to the extent that individual samples may show clear departures from equilibrium in the uranium series. This is especially possible since intermediate isotopes between ^{238}U and ^{226}Ra have half-lives up to 2.5 X 10^5 y (^{234}U) as shown in Table 2.1. This provides ample time for chemical changes and redistribution of elements to occur in near-surface soils.

Radium in the form of chlorides or other water soluble compounds is found in water supplies from lakes or streams to the extent of 0.5 to 50 Bq m^{-3}. Water drawn from wells will generally have a higher ^{226}Ra concentration than those from surface supplies. Sea water contains a relatively low concentration of about 4 Bq m^{-3}.

4.2.2 **Radium isotopes**

Sixteen isotopes of radium have been identified. Three of these (^{224}Ra, ^{226}Ra and ^{228}Ra) are widely known for their role in producing bone cancer in humans. The 224 and 228 isotopes are members of the naturally-occurring thorium series and the 226 isotope is in the uranium series.

Radium-224, the immediate precursor of thoron (^{220}Rn) in the thorium series (Table 2.2) is short-lived (3.64 d). In fact the thorium series as a whole contains relatively short-lived intermediate nuclides compared with the uranium series. The longest-lived member is ^{228}Ra, the immediate daughter of ^{232}Th at 5.8 y. As a result the members of the thorium series are found in soil and rocks essentially as expected from secular equilibrium with the long-lived parent ^{232}Th at 1.4 X 10^{10} y.

It is ^{226}Ra, in the uranium series that is the well known radium discovered by Madame Curie, the radium extensively used for medical purposes for many years, and the parent of the ubiquitous radon which has received wide notoriety in recent years as a possible danger in homes and work places.

The hazard of radium to humans was first learned when paint-laden brushes were made pointed by inserting in the mouths of workers applying luminous material to watch dials. Severe cases of "radium jaw" were observed up until the 1920's when tolerance doses and work guides were introduced.

4.3 EMANATION PROPERTIES

The effectiveness of radium in supplying radon to the soil pores for transport to the atmosphere depends not only upon the total concentration of radium atoms present per unit mass but also upon the fraction of those atoms in the soil and rock which are located in the matrix or on soil particle surfaces so that the newly formed radon atoms can escape into the pores and capillaries. The ratio of the radium atoms so situated to the total number of radium atoms present is called the "emanation coefficient." This fraction is also referred to as "emanating power" or "emanating fraction" in the literature. Values vary widely from 0.02 in lava fields in Hawaii to 0.76 in agricultural soils nearby (ref. 8).

The concept of "emanation coefficient" can best be illustrated by the surface-to-volume ratio for a given set of particles making up a soil. The diffusion coefficient for radon atoms in a solid is only about 10^{-7} of that for air. Hence, only atoms formed very near or on the surface of soil particles will reach the soil air space.

The transfer of radon atoms from mineral grains to soil capillaries is also dependent upon water films in the interstitial pore spaces. The recoil range of a radon atom at the instant of decay of the parent radium atom is about 60 μm in air, 0.006 μm in water, and even less within a mineral grain. If the average distance between soil particles is less than the recoil range of the radon atom in air, many of these recoiling atoms (or ions) will penetrate the surface of the neighboring soil grain and be lost to the soil pore space. On the other hand, if a film of water surrounds the grain, there is a higher probability that it will remain in the air space between soil particles (ref. 8). If too much water is present, soil transport procedures are impeded since the diffusion coefficient in water is of the order of 10^{-3} of its air value (ref. 9).

Mean values for ^{226}Ra, "effective ^{226}Ra" and the emanation coefficients for soil samples from New Mexico, Texas, Hawaii, Alaska and Iceland in the period 1969-1973 by the New Mexico Tech Radon Laboratory are given in Table 5.2. Samples were taken from the top 0.15 m of the soil and thoroughly dried before analysis. Errors for each site represent one standard deviation for the samples from that site. Overall mean values of 32 ± 15 Bq kg^{-1} (0.87 pg g^{-1}) with a range of 3.6 - 84 Bq kg^{-1} for ^{226}Ra, 12 ± 6 Bq kg^{-1} with a range of 0.7 - 34 Bq kg^{-1} for "effective ^{226}Ra" and a mean emanation coefficient of 0.36 were obtained for the sites indicated.

The highest ^{226}Ra concentration given in Table 5.2 is found in New Mexico as might be expected since the Grants Mineral Belt within the state has been one of the major sources of uranium ore in the world. On the other hand, island sources such as Hawaii show concentrations only about one-half as large. The ^{226}Ra present in the soils on the islands of Hawaii, however, has a greater chance of providing radon (^{222}Rn) to the soil gas and the atmosphere since the emanation coefficients are much larger than those of inland areas (New Mexico, Texas and Alaska).

4.4 RADON, THORON AND ACTINON

The chemical element radon with atomic weight 86 is the heaviest of the inert gases which include neon, argon, krypton and xenon as well. The common isotopes of radon are found in the environment because they are produced continuously by the decay of longer-lived nuclides found in minerals containing uranium, thorium, or actinium. A total of 20 isotopes of radon have been identified.

Radon (^{222}Rn) comes from ^{226}Ra in the uranium series, thoron (^{220}Rn) from ^{224}Ra in the thorium series and actinon (^{219}Rn) from ^{223}Ra in the actinium series. All three result from the alpha decay of the parent and all have relatively short half-lives: ^{222}Rn (3.82 d), ^{220}Rn (55.6 s) and ^{219}Rn (4.0 s) (Tables A.1, A.2 and A.3). Even though thorium (^{232}Th) is generally more abundant than uranium in the earth's crust, the probability for decay is smaller; hence, the production rates of ^{222}Rn and ^{220}Rn in the soil are roughly the same. Much of the thoron (^{220}Rn) decays before reaching the earth's surface due to its short half-life of 52 s. Actinon (^{219}Rn) with a half-life of only 1.92 s has little chance of escaping from the soil by diffusion because of an even shorter half-life than thoron (^{220}Rn). As a result actinon is extremely rare in the atmosphere.

There are several characteristics of radon and thoron that account for the dominance of these isotopes, and their short-lived daughters in the dose to the human respiratory tract. These factors include (1) the wide spread availability of the parent nuclides, ^{226}Ra and ^{224}Ra, in the rocks and soil and building materials, (2) the transport processes available for radon as an inert gas, (3) the significant list of short-lived daughter products as given in Table 4.2, and (4) an alpha energy in each series of approximately 25 MeV per decaying radon or thoron atom. The result is that inhaled radon progeny account for 55% of the annual dose to the human lung compared with nearly equal doses of about 15% each from the other natural ionizing radiations including cosmic rays, terrestrial radiation and natural radionuclides such as ^{40}K in the body (ref. 10).

4.4.1 Radon (^{222}Rn)

Radon or "radium emanation" as it was first called on its discovery by Dorn in 1900 was recognized as a unique chemical element having a high level of penetrating radiation associated with it.

TABLE 4.2

Physical properties of radon (^{222}Rn) and thoron (^{220}Rn) and their short-lived decay products. From CEC 1987 (ref. 11).

RADON-222

Radionuclide and Historical Symbol	Half Life	No. Atoms Per Bq	Principal Radiation(s)	Energies (MeV) and Intensities
^{226}Ra(Ra)	1,600 y	7.4 X 10^{10}	α	4.8 (95%)
^{222}Rn(Rn)	3.82 d	4.8 X 10^{5}	α	5.5 (100%)
^{218}Po(RaA)*	3.11 m	2.6 X 10^{2}	α	6.0 (~100%)
^{214}Pb(RaB)*	26.8 m	2.3 X 10^{3}	β γ	0.65 (50%) 0.35 (36%)
^{214}Bi(RaC)	19.7 m	1.7 X 10^{3}	β γ	3.3 (19%) 0.61 (47%)
^{214}Po(RaC')	164 μs	2.4 X 10^{-4}	α	7.7 (100%)
^{210}Pb(RaD)	21 y	9.6 X 10^{8}	β	0.016 (85%)

Branch Decays of less than 0.02% are neglected.

THORON-220

	Radionuclide and Historical Symbol	Half Life	No. Atoms Per Bq	Principal Radiation(s)	Energies (MeV) and Intensities
	^{224}Ra(Ra)	3.64 d	4.5 X 10^{5}	α	5.68 (94%)
	^{220}Rn(Tn)	55 s	7.9 X 10^{1}	α	6.29 (100%
	^{216}Po(ThA)	0.15 s	2.2 X 10^{-1}	α	6.78 (100%)
	^{212}Pb(ThB)	10.6 h	5.5 X 10^{4}	β	0.35 (81%)
64%	^{212}Bi(ThC)	60.6 m	5.2 X 10^{3}	β	2.27 (54%)
	^{212}Po(ThC)	0.304 μs	4.3 X 10-7	α	8.78 (100%)
	^{208}Pb(ThD)	Stable			
or 36%	^{212}Bi(ThC)	60.6 m	5.2 X 10^{3}	α	6.05 (70%)
	^{208}Tl(ThC")	3.1 m	2.7 X 10^{2}	β	1.8 (50%)
	^{208}Pb(ThD)	Stable		γ	0.58 (86%)

Details of the decay chain of ^{222}Rn are given in Table 4.2. There are four short-lived daughters in the ^{223}Rn series - two alpha and two beta-gamma emitters. The longest lived member is ^{214}Pb with a half-life of 26.8 m. Not shown beyond the 21-year ^{210}Pb are ^{210}Bi (5.0d, β) and ^{210}Po (138 d, α) before this series ends at stable ^{206}Pb. From the standpoint of potential damage to the human lung the alpha decays of ^{218}Po and ^{214}Po totalling 13.7 MeV of energy are of chief concern. These nuclides are deposited in the lung through the breathing of air containing ^{218}Po, ^{214}Pb, and ^{214}Bi attached to dust or in free-ion or atomic form.

Of the radon isotopes, ^{222}Rn, a decay product of ^{226}Ra in soil or building materials, with its short-lived decay products constitutes the major concern as a health hazard in the environment.

4.4.2 Thoron (^{220}Rn)

The discovery of thoron is credited to Sir Ernest Rutherford of Cambridge University. He reported his finding of "thorium emanation" and its "active deposit" in the January-February 1900 issue of the Philosophical Magazine.

Thoron (^{220}Rn) has a half-life of only 55 s and decays with the emission of an alpha particle having an energy of 6.29 MeV. The decay chain is given in Table 4.2. The longest lived daughter is the 10.6 h ^{212}Pb which shows up in a decay series of radon and thoron decay products on atmospheric dust from a typical indoor/outdoor atmosphere (Fig. 4.1). The short-lived decay series has two alphas: ^{216}Po, 6.78 MeV and ^{212}Po, 8.78 MeV (or ^{212}Bi, 6.05 MeV) comparable to the two alphas in the radon (^{222}Rn) chain. In addition there are two beta emissions and some gamma rays. Extensive studies of thoron and its decay products have been made in Schery (ref. 9), Israel (ref. 12), Crozier and Biles (ref. 13), and others.

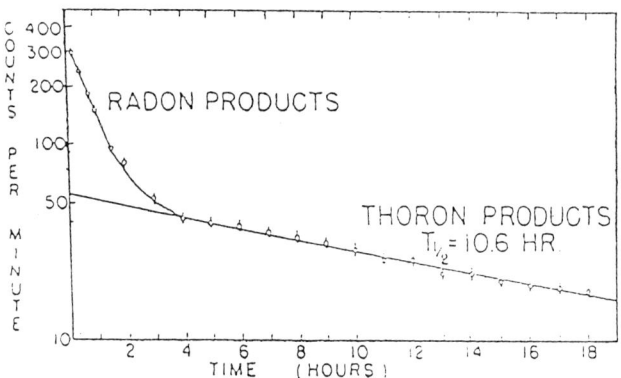

Fig. 4.1 Decay curve of alpha activity from a sample of airborne dust. Composite characteristics of the decay products of radon (^{222}Rn) having an effective half-life of about 40 minutes and thoron (^{220}Rn) dominated by the 10.6 hour half-life of ^{212}Pb (ThB) are clearly indicated. Collection time was three hours.

4.4.3 Actinon (^{219}Rn)

This isotope originates from the alpha decay of ^{223}Ra in the actinium series. With a half-life of only 3.92 sec, ^{219}Rn has little time for diffusion to the atmosphere from soil and rocks. In addition the head of the actinium series, ^{235}U, has an abundance in natural uranium of only 0.71% by weight. It is extremely rare in the atmosphere and as a result ^{219}Rn can generally be neglected in radon dose considerations.

Now that the occurrence and abundance of the parent elements of ^{220}Rn and ^{222}Rn have been reviewed, their transport from soil to air will be described in the following chapter.

REFERENCES

1 CRC, Handbook of Chemistry and Physics, 60th edn., R.C. Weast (Ed.) and M.J. Astle (Assoc. Ed.) Chemical Rubber Publishing Co., Boca Raton, FL, 1979, p. B-25.
2 M. Eisenbud, Environmental Radioactivity, 3rd edn., Academic Press Inc., 1987, pp. 149-152.
3 NCRP 94, Exposure of the Population of the United States and Canada from Natural Background Radiation, National Council on Radiation Protection and Measurements, Bethesda, MD, 1987, p. 61.

4   R.L. Kathren, Radioactivity in the Environment, Harwood, NY, 1984, pp. 64-66.
5   NCRP 97, Measurement of Radon and Radon Daughters in Air, National Council on Radiation Protection and Measurements, Bethesda, MD, 1988, pp. 28-29.
6   P. Ziqiang, Y. Yin and G. Mingqiang, Natural radiation and radioactivity in China, Radiation Protect. Dosim., 24(1-4), (1988) 29-38.
7   T.E. Myrick, B.A. Berven and F.F. Haywood, Determination of concentrations of selected radionuclides in surface soil in the U.S., Health Phys., 45, (1983) 631-642.
8   M.H. Wilkening, Radon from the island of Hawaii, Science, 183, (1974) 413-415.
9   M. Wilkening, Radon transport in soil and its relation to indoor radioactivity, Science of the Total Environment, 45, (1985) 219-226.
10 DOE/ER-0405 RADON, FY-1988 Radon Research Program, United States Department of Energy, Office of Health and Environmental Research, Washington, DC, 1989, p. 85.
11 CEC Radiation Protection, Exposure to Natural Radiation in Dwellings in the European Communities, Commission of the European Communities, Luxembourg, 1987, pp. 111-112.
12 A.W. Israel, Thoron (Rn-220) measurements in the atmosphere and their applications in meteorology, Tellus, 17, (1965) 383-388.
13 W.D. Crozier and N. Biles, Measurement of radon-220 (thoron) in the atmosphere below 50 centimeters, J. Geophys. Res., 71(20), (1966) 4735-4741.

CHAPTER 5

RADON – SOIL TO AIR

Concentrations of uranium-238 and thorium-232 in soils are approximately equal at about 40 Bq kg^{-1} (Table 4.1). Barring any unusual removal of intermediate members of the decay chains from the soil, it is reasonable to expect that the production rates of ^{222}Rn and ^{220}Rn in the soil will be roughly equal also.

5.1 TRANSPORT PROCESSES

The transport of the radon isotopes from the soil to the air is treated in this chapter. Knowledge of the processes involved is of importance in the determination of radon levels in houses, uranium exploration, earthquake prediction, and a more complete understanding of the natural radiation environment.

Processes involved in the transport of radon from the soil to the surface are related to the size and configuration of the spaces occupied by the soil gas. These spaces may vary from molecular interstices to large underground caverns. The openings may be isolated, interconnected, or dead end. The pore volume may be a small or substantial fraction of the gross soil volume. All these characteristics are important to the radon transport problem.

The near-surface transport processes can be broken down into two major categories: (1) a microscopic process in which diffusive and viscous flows are dominant and (2) a macroscopic process in which flow occurs through cracks, fissures, and underground channels. A key reference for radon transport in the soil is Tanner (ref. 1).

5.1.1 Diffusion and viscous flow

The use of the kinetic theory of gases to transport processes in soil dates back to the early part of this century. Basically a constituent of the soil gas such as radon flows in a direction opposite to that of the increasing concentration gradient (Fick's Law). If the diffusive flow is upward in the z-direction, $J_d = -D(dC/dz)$ where J_d is the diffusion current density, D is the bulk diffusion coefficient, and C is the concentration of ^{222}Rn in the interstitial space. If there is a fluid current density J_f also in the positive z-direction, then $J_f = Cv$, where v is the transport velocity. The steady-state equation for a semi-infinite, homogeneous, porous medium is

$$\frac{D}{\epsilon} \frac{d^2C}{dz^2} - \frac{1}{\epsilon} \frac{d(vC)}{dz} - \lambda_{Rn} C + \phi = 0 \qquad (5.1)$$

where ϵ is the soil porosity, λ_{Rn} is the decay constant of ^{222}Rn, and ϕ is the production rate per unit volume of ^{222}Rn atoms that are free to migrate in the interstitial space. The terms in eqn. (5.1) represent, in order, diffusion, flow, decay, and production. For flow by molecular diffusion alone, and with boundary conditions $C = 0$ at $z = 0$ and $C = \phi/\lambda_{Rn}$ at $z = -\infty$, the following solution of eqn. (5.1) is obtained:

$$C = \frac{\phi}{\lambda_{Rn}} \left\{ 1 - \exp\left[\left(\frac{\epsilon \lambda_{Rn}}{D}\right)^{\frac{1}{2}}\right]\right\} . \qquad (5.2)$$

This shows that the concentration increases with depth reaching to within 3% of the concentration at infinite depth at 5 m when numerical values from ref. 2 are used. Molecular diffusion as a significant transport mechanism is limited to the first few meters of surface soil.

Once the solution of eqn. (5.1) is found for a specified form of the transport velocity v, the total ^{222}Rn current density J in the soil is

$$J = -D \frac{dC}{dz} + Cv \qquad (5.3)$$

The dependence of v on z in fluid flow is found by application of Darcy's Law $v = -(k'/\eta)dp/dz$ where p is the pressure, k' is the soil permeability, and η is the dynamic viscosity of the fluid. Equations (5.1), and (5.3) furnish the points of departure for models relating the physical properties of soils to transport that is the result of both molecular diffusion and a pressure-induced flow. Variations of this simple model have been applied to ^{222}Rn in near-surface soils in Clements and Wilkening (ref. 2), Alekseev et al. (ref. 3), Kraner, Schroeder, and Evans (ref. 4) and others.

No complete mathematical description of flow is possible because of the great complexity and diversity of the details of soil structure. However, the great technological importance of soil physics in agriculture and the flow and recovery of groundwater in porous media has resulted in great progress being made in recent years toward the development of mathematical models relating the physical properties of porous media.

An approach that is used widely is one in which the interconnected pores are treated as straight nonintersecting cylindrical tubes or capillaries (ref. 5). The diameters are given by an appropriate distribution function for a given porous medium (ref. 6). Wakao, Otani, and Smith (ref. 7) have derived equations for flow and diffusion in capillaries covering the complete range of flow from Knudsen to Poiseuille. The characteristics of these flows will be reviewed in order of increasing size of the equivalent capillary radius, r. A summary is given in Table 5.1 (ref. 8).

TABLE 5.1

Transport in soils. From Wilkening (ref. 8).

	Knudsen diffusion	Ordinary bulk diffusion	Poiseuille flow (laminar/viscous)
Soil	Compact, fine-grained	Porous	Fine cracks, fissures
Equivalent capillary radius	$r/\lambda < 0.1$ ($r < 0.01$ μm)	$r/\lambda > 10$ ($r < 1$ μm)	$r > \lambda$ ($r > 0.001$ mm.)
Flux	$J = -D \frac{dC}{dz}$	$J = -D \frac{dC}{dz}$	$F = \frac{\pi r^4 \bar{p}}{8\eta kT} \frac{dp}{dz}$
Coefficient	$D = \frac{2}{3} r\bar{v}$	$D = \frac{\bar{v}}{3}$	$F = A\, r^4 \bar{p} T^{-1} \eta^{-1} \frac{dp}{dz}$
	$= A\, rT^{\frac{1}{2}} M^{-\frac{1}{2}}$	$= A\, T^{\frac{3}{2}} p^{-1} M^{-\frac{1}{2}}$	

r = radius of capillary
λ = mean free path, 0.1 μm at STP
J = flux
D = diffusion coefficient
C = concentration of radon atoms
z = coordinate perpendicular to soil surface
A = constant
k = Boltzmann constant
F = flow, molecules per second
$\bar{v}$ = mean molecular velocity
T = temperature, °K
M = atomic/molecular mass
p = pressure
η = dynamic viscosity

(i.) <u>Knudsen diffusion in fine capillaries</u>.

This type of transport occurs in capillaries having diameters that are small compared with the mean free path of the gas molecules. Since the mean free path in air at STP is about 0.1 μm, the capillary would have to have a radius of the order of 0.01 μm. Only the smallest colloidal clays have constituent particles in this size range. <u>Diffusion in fine capillaries (Knudsen) is not effective in radon transport in soils near the earth's surface</u> since pore spaces in most soils are much too coarse grained to fit this capillary model.

Although Knudsen diffusion does not play an important role at normal pressures on the earth, it provides the predominant type of diffusive transport on the moon where the ^{222}Rn atoms collide almost exclusively with solid material within the lunar regolith. Diffusion of radon and other inert gases in the lunar crust has been considered by Kraner et al. (ref. 4).

(i.i.) *Ordinary bulk molecular diffusion*.

This mode of transport with or without the addition of pressure-induced flow is the primary mechanism for radon transport in soil. An analysis of the dependence of the flow on physical variables within the equivalent-capillary concept is instructive. Table 5.1 shows that the bulk diffusion coefficient D for a constituent such as ^{222}Rn in the soil gas is dependent upon the mean free path of ^{222}Rn in the pores and the mean molecular velocity. On substitution from kinetic theory, the diffusive flux is seen to be proportional to $T^{\frac{3}{2}}$, p^{-1}, and $M^{-\frac{1}{2}}$. It is independent of the equivalent capillary radius, r. Note that ^{222}Rn with its large atomic mass will diffuse only about one-third as rapidly as other constituents of the soil gas since the flow in ordinary molecular diffusion is inversely proportional to the square root of the atomic or molecular mass (ref. 8).

An increase of temperature of 20°C results in an increase of J by only about 4% if all other factors remain the same. The important role played by molecular diffusion in the soil as a transport mechanism near the earth's surface is reviewed by Tanner (ref. 1). He points out that the diffusion length, the average distance an atom can move through dry soil before decaying is about 1.6 m for ^{222}Rn but only 2 cm for ^{220}Rn. Hence, much of the ^{220}Rn decays before reaching the earth's surface due to its short half-life of 55 sec.

(i.i.i.) *Poiseuille flow*.

This type of flow occurs in long straight capillaries of circular cross section. Again the diameter must be large compared with the mean free path. The velocity of transport is small (low Reynolds number), with the result that the flow is laminar and dependent upon the viscosity of the gas. This type of flow can be a major contributor to radon transport in a variety of crustal materials where pressure gradients are available and where this type of capillary is an appropriate approximation to the physical structure.

It is important to remember that the diffusion or flow equations given in Table 5.1 when taken separately are not adequate for explaining transport in a particular medium. A model combining all flows given might be necessary. Further, the transition from Knudsen to ordinary bulk diffusive flow and the "slip" flow term for viscous flow in cylindrical capillaries have been omitted. See ref. 9 for detailed treatment of combinations of these flows. In general diffusion and flow mechanisms of the type indicated are responsible for a major portion of the upward transport of radon from soil depths down to several meters below the surface.

Much work is needed to quantify the changes observed in radon flux due to wind-scour and moisture effects. Wind-induced transport includes both the direct transfer of momentum on exposed surfaces from surface air to the soil gas, and the effects of the alternating pressure component induced in near-surface layers of the soil originating from turbulent airflow across the surface. The latter problem has been studied both by soil scientists interested in water-vapor transport (ref. 10) and from the viewpoint of seismometer responses to small atmospheric pressure cells as they propagate along the surface at mean wind velocities (ref. 11).

5.1.2 Flow in channels

Transport of radon-rich air from inside the earth's crust is known to occur through openings and fissures in the ground that are much larger than those treated above. Radon-222 moves through the soil gas by ordinary diffusion over distances of only a few meters. Pressure-induced Darcy velocities of the order of 10^{-6} m/sec are sufficient to allow ^{222}Rn to move only about one-third of a meter in one half-life.

Situations do occur where transport over many tens or hundreds of meters can take place. There are two main cases to be considered; the first involves the transport of ^{222}Rn by air movement through cracks, fissures, or underground openings where pressure gradients exist, and the second requires temperature differences favorable to vertical convective transport through relatively large openings. The first case has been illustrated by ^{222}Rn concentration and air flow measurements in a horizontal tunnel some 172 m long in a hillside (ref. 12). The net flow of air into the tunnel each day apparently displaced an equal volume of radon-rich air out of the tunnel. The result is that approximately 200 Bq/sec of ^{222}Rn is added to the air above the hill in excess of the normal exhalation by diffusion. This is equivalent to the quantity of ^{222}Rn from approximately 8,000 m^2 of surface in the same area.

In the case of the Carlsbad Caverns (ref. 13) the temperature in the cave is lower than the outdoor average temperature during the summer months. The vertical air column through the entrance is stable and little or no exchange takes place. During the winter months the outdoor temperature is lower than that in the cave. Warm air from the interior of the cave rises bringing radon-rich air with it. This convective exchange results in about 6,300 Bq/sec of radon being fed to the atmosphere averaged over a year. This amount of radon exhalation is about what might be expected from 0.25 km^2 of surface soil.

Volcanoes as sources of atmospheric radon have been investigated in Hawaii, Iceland and Japan. Although the fumaroles and plumes sampled show radon levels well above that of the ambient atmosphere, the overall contribution of volcanoes was found to be negligible (ref. 14).

5.2 MEASUREMENT

The transfer rate per unit area of radon from the Earth or any solid substances to the atmosphere is referred to as the radon flux density or exhalation rate. In SI units it is Bq m^{-2}s^{-1}. Other units used have included atoms cm^{-2} s^{-1} and Ci m^{-2}s^{-1}. The flux density gives a measure of the source strength.

The radon flux density from a given material can be measured in the laboratory by placing a sample in an enclosed chamber and allowing the radon to build up after first replacing the air in the container with radon-free air or nitrogen. Field measurements are emphasized in the methods illustrated in Fig. 5.1.

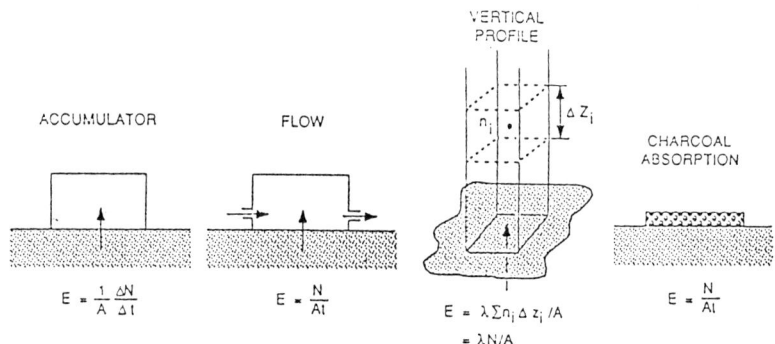

Fig. 5.1 Methods used in measuring radon flux. E is the flux, A is the area covered by the device or system, t is the accumulation time, N is the quantity of radon and λ is the decay constant for radon.

5.2.1 Accumulator

One of the earliest and most widely used devices for measurement of radon flux is simply a closed container resting on the soil surface. See Fig. 5.2. The device used by Wilkening et al. (ref. 15) in their early work consisted of an aluminum ring about 0.5m in diameter that could be placed in the soil with the main accumulator container on top. The total accumulator volume

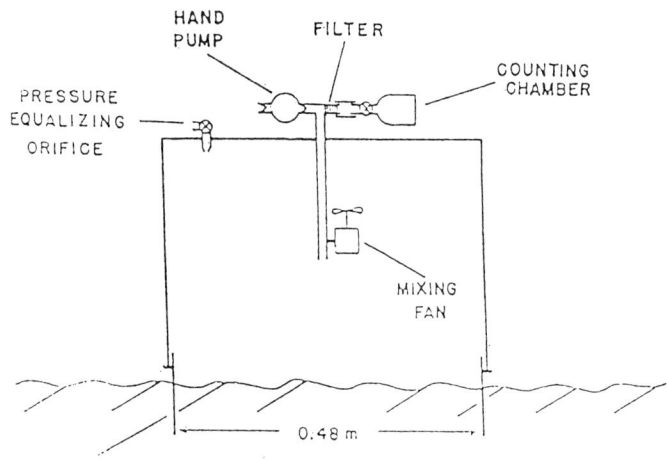

Fig. 5.2 Accumulator for measurement of radon flux density (From Wilkening et al., 1972).

was approximately one-third of a cubic meter. Each unit is equipped with a sampling tube, pressure equalizing orifice, and a small mixing fan. Samples are taken in an evacuated 100 ml alpha scintillation chamber after running the fan for one minute to insure complete mixing.

The time rate of increase in the number of ^{222}Rn atoms ($\Delta N/\Delta t$) in the accumulator having an area A is directly related to the exhalation rate (flux) E by

$$E = \frac{1}{A} \frac{\Delta N}{\Delta t} \tag{5.4}$$

provided (1) the accumulating time is short compared to the half-life of radon-222, (2) the concentration in the accumulator remains low enough to prevent appreciable back diffusion, and (3) the presence of the accumulator does not seriously disturb the exhalation process.

Other early measurements of flux density using the accumulator method were made by Kraner et al. (ref. 16) and Sisigina (ref. 17). An adaptation of this method using the electrostatic collection of the decay-product positive ions has been described by Rosen (ref. 18).

5.2.2 Flow method

This method, first described by Pearson and Jones (ref. 19), provides for circulation of sample air through the accumulation vessel at a low flow rate and thence through a charcoal trap which collects the radon. An advantage of this method is that the concentration gradient does not build up in the measurement vessel. Another advantage is that the intake air flow can resemble natural conditions with one side of the intake vessel being exposed to the soil surface.

Modifications of the flow system have been developed with particular attention being given to detectors used. More recently Schery and Gaeddert (ref. 20) have developed a unit which continuously monitors the radon in the outlet airstream.

Another flow-through device has been described by Watnick et al. (ref. 21) who used a hemispherical container placed on the ground from which air was withdrawn. While this system can operate continuously with good reliability, it does apply negative pressure to the soil surface being measured which may disturb the equilibrium flux rate.

5.2.3 Adsorption method

In this case activated charcoal is placed directly on the area to be measured. A gauze base is used to contain the charcoal and facilitate its recovery. After exposure for a number of hours, the charcoal is placed in a counting chamber where the gamma rays from the ^{214}Bi radon daughters are measured. The thoron (^{220}Rn) exhalation can also be estimated by counting the ^{208}Tl daughters. Megumi and Mamuso (ref. 22) have used this method while a simpler technique involving the use of charcoal gas mask canisters has been described by Countess (ref. 23). The canister procedure is simple, and the sampler can be readily sealed to vertical surfaces with modeling clay or flexible caulking material. On the other hand, it requires calibration against more conventional accumulators and the degree to which it perturbs the system being measured is not known.

5.2.4 Vertical profile

The basic assumption for this method is that the total quantity of radon in a vertical column of air of fixed area represents that which has been exhaled by that same area of soil under equilibrium conditions. This procedure has been used in the Soviet Union (ref. 24-25) and by Wilkening et al. (ref. 15) in the United States. The fact that aircraft or balloons are required for sampling has limited the use of this system for the measurement of vertical profiles. In addition, strong horizontal concentration gradients are known to exist particularly where maritime and continental air masses interact; hence, this approach is applicable only where air mass systems have long trajectories over continents.

The flux (E) is calculated from $EA = dN/dt$ where A is the base area of the column and N is the total number of radon atoms in the column. Taking $A = 1 \text{ m}^2$, $dN/dt = \lambda N$, and λ the decay constant of ^{222}Rn, $E = \lambda N$. N can be determined by summing the vertical distribution of concentration, n:

$$N = \sum_t n_i \Delta z_i \tag{5.4}$$

where n_i is the concentration in the ith interval of vertical extent, Δz_i. In summing the $n_i \Delta z_i$ terms, the structure of the atmosphere with respect to the dominant layers is taken into account. Air samples are taken from ground level through the mixing layers by means of captive balloons or aircraft. Average values for ^{222}Rn flux can be determined for large areas by this method; however, the condition, of no advective change is a restricting one.

5.2.5 Soil concentration gradient

As noted in Section 5.1.1. above the transport of radon from soil to atmosphere is the result of concentration and pressure gradients. If measurements can be made under stable pressure conditions and when rain and wind do not interfere, it is possible to establish the soil concentration gradient and to estimate the corresponding flux. As Clements and Wilkening (ref. 26) showed, this requires an independent measurement of the diffusion constant of radon in soil as well as determinations of the soil radon concentration, as the soil-air interface is approached. The method has been applied by Mochizuki and Sekikawa (ref. 27).

Schery et al. (ref. 28) used concentrations of ^{210}Pb in the surface layers of soil as an indication of the time-averaged gradient of radon concentration (^{210}Pb is the long-lived member of the ^{222}Rn series). Their results for flux rate agree with other measurements made with closed and flow-type accumulators. Fleischer (ref. 29) described another method that uses solid state nuclear track detector techniques in determining soil radon gradients. His monitors were placed in cups at different depths and radon concentrations were inferred from track densities. The problem involved in applying Fickian diffusion only to radon transport must be resolved in this case. The use of alpha-sensitive materials and detectors is being investigated in other laboratories.

A number of methods are available for measuring radon flux densities as described above. A problem does exist in that the field techniques described do affect the system being measured. An ideal method would not disturb the surface being measured nor would it interfere with meteorological effects such as those due to rain or wind. Much needs to be done before the real spatial and temporal variations in radon flux can be understood in detail.

TABLE 5.2

Radium-226 concentrations, emanation coefficients, and ^{222}Rn flux densities for various locations. From references 15, 18, and 32.

Location	Soil	Number Sites	Number Samples	^{226}Ra (Bq kg^{-1})	^{226}Ra-eff (Bq kg^{-1})	Em Coef.	^{226}Rn flux density (mBq m^{-2}s^{-1})
Alaska - Fairbanks	Silt and sandy loam	4	14	36 ± 3	10 ± 3	0.28	5
Hawaii	Deep volcanic	12	29	26 ± 18	18 ± 20	0.69	32
New Mexico	Desertic, gravel and sandy loams	15	165	59 ± 25	12 ± 6	0.20	30
Texas	Clay-loam, coastal to high plains	3	9	33 ± 14	14 ± 11	0.42	10
TOTAL/MEAN		34	217	38 ± 14	14 ± 3	0.40 ± 0.22	19 ± 14

5.3 RESULTS OF FLUX DENSITY MEASUREMENTS

The needs for flux density measurements are several. Chief among them is the direct identification of the source of radon which may lead to health concerns in a residence, school, or working place. The large flux of radon from the walls of a uranium mine coupled with poor ventilation produced levels of radon which resulted in the death of a number of miners from lung cancer. Radon in houses can reach levels of concern due to the flux from soil and rocks below the floors. Diagnostic measurements are needed for remedial action.

A knowledge of radon flux can be a necessary part of scientific studies aimed at a more complete understanding of radon and radon decay product concentrations in the environment. The results of measurement of radon flux density in a variety of locations and over a range of soil types are given in Table 5.2. The mean ^{226}Ra concentrations for the soils in the table average 38 ± 13 Bq/kg (1.05 ± 0.52 pg/g) compared with a typical value of 40 Bq kg^{-1} for rocks and soils (ref. 30). The mean value for ^{226}Ra noted above and an average ^{226}Ra-effective of 14 ± 6 Bq/kg yield an emanation coefficient of 0.37. This can be compared with those listed by Nazaroff and Nero (ref. 31) which show a mean emanation coefficient of 0.34 for some 42 soil samples.

The mean ^{222}Rn flux density from Table 5.2 is 19 ± 12 mBq m^{-2}s^{-1} showing good agreement with a figure of 17 mBq m^{-2}s^{-1} for continental soils given in NCRP No. 97 (ref. 30). Radon-222 flux is quite variable as noted by the large standard deviation. This is expected because of the wide variation in effective ^{226}Ra concentrations, distribution of ^{226}Ra atoms on and within soil and rock particles, soil porosity, grain size, moisture content, pressure differentials, and other factors. From this knowledge of flux density one can see that the decay of the ^{226}Ra in soil within a meter or so of the surface results in a net flux of about 20 m Bq m^{-2}s^{-1}. For a typical house covering some 300 m^2 the availability of ^{222}Rn would be 6,000 m Bq s^{-1}.

Based upon a knowledge of the sources of radon in the rocks and soils of the earth's crust and the mechanisms of transport from soil to air, attention is now directed to quantitative analyses of radon and thoron concentrations in the atmosphere in the next chapter.

REFERENCES

1   A.B. Tanner, Radon migration in the ground: a supplementary review, in: Natural Radiation Environment III, Technical Information Center/U.S. Department of Energy, Springfield, VA, 1980, pp. 5-56.
2   W.E. Clements and M.H. Wilkening, Atmospheric pressure effects on ^{222}Rn transport across the earth-air interface, J. Geophys. Res., 79, (1974) 5025-5029.

3   V.V. Alekseev, A.G. Grammakov, A.I. Nikonov and G.P. Tafeev, Radiometric Methods in the Prospecting and Exploration of Uranium Ores, USAEC Report AEC-tr-3738, Book 2, pp. 429-438, translated from Radiometricheskie Metody Poiskov i Razvedki Uronovykh Rud, State Scientific-Technical Publishers of Literature on Geology and Mineral Resources Conservation, Moscow, USSR, 1957.
4   H.W. Kraner, G.L. Schroeder and R.D. Evans, Measurements of the effects of atmospheric variables on ^{222}Rn flux and soil-gas concentrations, in: J.A.S. Adams and W.M. Lowder (Eds.), The Natural Radiation Environment, Univ. of Chicago Press, Chicago, IL, 1969, pp. 191-215.
5   A.E. Scheidegger and K.H. Liao, Fundamentals of transport phenomena in porous media, Elsevier Publishing Company, Amsterdam, 1972.
6   F.A.L. Dullien and V.K. Batra, Determination of the Structure of Porous Media, in: Flow Through Porous Media, American Chemical Society Publications, Washington, DC, 1970.
7   N. Wakao, S. Otani and J.M. Smith, Significance of pressure gradients in porous materials: Diffusion and flow in fine capillaries, AIChE J., 11: 435-439, 1965.
8   M.H. Wilkening, Radon transport processes below the earth's surface, in: T.F. Gesell and W.M. Lowder (Eds.) Natural Radiation Environment III, Technical Information Center/U.S. Department of Energy, Springfield, VA, 1980, p. 93.
9   G.R. Youngquist, Diffusion and flow of gases in porous solids, in: Flow Through Porous Media, American Chemical Society Publications, Washington, DC, pp. 58-69, 1970.
10 H. Fukuda, Air and vapor movement in soil due to wind gustiness, Soil Sci., 79, (1955) 249-256.
11 J. Peterson and N.A. Orsini, Seismic research observatories: Upgrading the worldwide seismic data network, EOS, Trans., Am. Geophys. Union, 57, (1976) 548-556.
12 M.H. Wilkening, Free air exchange in an open mine tunnel determined by radon measurements, J. Geophys. Res., 67, (1962), 2525.
13 M.H. Wilkening and D. Watkins, Air exchange and ^{222}Rn concentrations in the Carlsbad Caverns, Health Phys., 31, (1976) 139-145.
14 M.H. Wilkening, Radon 222 from the island of Hawaii: Deep soils are more important than lava fields or volcanoes, Science, 183, (1974) 413-415.
15 M. Wilkening, W.E. Clements and D. Stanley, Radon-222 flux measurements in widely separated regions, in: The Natural Radiation Environment II, J.A.S. Adams, W.M. Lowder and T.F. Gesell (Eds.), USAEC Report CONF-720805-P2 (National Technical Information Service), Springfield, VA, 1972, p. 717.
16 H.W. Kraner, G.L. Schroeder and R.D. Evans, Measurements of the effects of atmospheric variables on radon-222 flux and soil-gas concentration, in: The Natural Radiation Environment, (University of Chicago Press, Chicago), 1964, p. 191.

17 T.I. Sisigina, Radon emanation from the surface of some types of soils of the European part of the USSR and Kazakhstan, in Radioactive Isotopes in the Atmosphere and Their Use in Meteorology, I.L. Karol et al. (Eds.), English translation by USAEC, Washington, DC, 1967, pp. 29-33.

18 R. Rosen, Note on some observations of radon and thoron exhalation from the ground, New Zealand J. Sci. Technol., B38, (1956) 644-654.

19 J.E. Pearson and G.E. Jones, Emanation of radon-222 from soils and its use as tracer, J. Geophys. Res., 70, (1965) 5279.

20 S.D. Schery and D.H. Gaeddert, Measurements of the effects of cyclic atmospheric pressure variation on the flux of ^{222}Rn from the soil, J. Geophys. Res., Letters, 9, (1982) 835.

21 S. Watnick, N. Latner and R.T. Graveson, RADEX: An active monitor for continuous measurement of ^{222}Rn flux from the soil, USDOE Report EML-428 (Environmental Measurements Laboratory, New York), 1984.

22 K. Megumi and T. Mamuro, A method for measuring radon and thoron exhalation from the ground, J. Geophys. Res., 77, (1972) 3052.

23 R.J. Countess, Rn-222 flux measurement with a charcoal canister, Health Phys., 31, (1976) 455.

24 S.G. Malakhov and P.G. Chernysheva, On the seasonal variation in the concentrations of radon and thoron in the surface layer of the atmosphere, in: Report USAEC-tr6711 (Health and Safety Laboratory, New York), 1967, p. 60.

25 L.V. Kirichenko, Radon exhalation from vast areas according to vertical distribution of its short-lived decay product, J. Geophys. Res. 75, (1970) 3539.

26 W.E. Clements and M. Wilkening, Atmospheric pressure effects on ^{222}Rn transport across the air-earth interface, J. Geophys. Res., 79, (1974) 5025.

27 S. Mochizuki and T. Sekikawa, ^{222}Rn exhalation and its variation in soil air, in: Natural Radiation Environment III, T.F. Gesell and W.M. Lowder (Eds.), USDOE Report CONF-780422 (National Technical Information Service, Springfield, VA), (1980) 105.

28 S.D. Schery, D.H. Gaeddert and M.H. Wilkening, Factors affecting exhalation of radon from a gravelly sandy loam, J. Geophys. Res., 89, (1984), 7299.

29 R.L. Fleischer, Radon flux from the earth: Methods of measurement by the nuclear track technique, J. Geophys. Res., 85, (1980) 7553.

30 NCRP 97, Measurement of Radon and Radon Daughters in Air, National Council on Radiation Protection and Measurements, Bethesda, MD, 1988, pp. 28-29.

31 W.W. Nazaroff and A.V. Nero, Jr., Radon and its decay products in air, John Wiley and sons, NY, 1988, pp. 73-79.

32 M. Wilkening, Radon transport in soil and its relation to indoor radioactivity, Sci. Total Environ., 45, (1985) 219-226.

CHAPTER 6

RADON IN THE ATMOSPHERE

The occurrence of radon in the atmosphere has been known since soon after it was discovered in 1900. However, its distribution in space and time, its physical state as an atmospheric component, and its mode of participation in the dynamics of the atmosphere are not yet fully understood.

6.1 GROUND LEVEL

The ^{222}Ra distributed in the soil and rocks of the Earth's crust produces ^{222}Rn, its daughter product, at a constant rate of approximately 40 Bq kg^{-1}. The ^{222}Ra-effective, that which is located on soil or rock particle grains in a position favorable for the escape of ^{222}Rn atoms to the soil gas in the capillaries is then 16 Bq kg^{-1} (for a typical emanation coefficient of 0.4). The ^{222}Rn atoms produced in this manner are free to diffuse in the air space between the soil particles where levels of ^{222}Rn in the soil air are of the order of 55,000 Bq m^{-3} indicative of ^{222}Rn concentrations at three meters or more below the surface (ref. 1).

As the soil-air interface is approached from below the concentration decreases but may still be about 1,000 Bq m^{-3} only 0.05m (5 cm) below the surface. See Schery et al. (ref. 2). Since typical outdoor air ^{222}Rn concentrations are about 8 Bq m^{-3} (ref. 3), at one meter above ground level the change in concentration across the soil-air interface is a factor of 100 or more. Based upon a diffusion coefficient of 0.05 cm^2 s^{-1} (ref. 4) for ^{222}Rn in soil it can be shown that the concentration at a depth of 1 meter has a concentration that is one-half of that at the surface. On the other hand the height in the atmosphere at which the concentration reduces to one-half that at ground level is about 1,000 m based upon a diffusion coefficient in the atmosphere of 5 X 10^4 cm^{-2} s^{-1} (ref. 5). The corresponding figures for ^{220}Rn (thoron) are a depth of 1.5 cm and a height of 14 cm.

The major discontinuities in ^{222}Rn and ^{220}Rn concentrations at the earth's surface are responsible for flux densities from earth-to-atmosphere of about 16 mBq m^{-2}s^{-1} for ^{222}Rn and about 1.5 Bq m^{-2} s^{-1} for ^{220}Rn. It is this transport of the radon isotopes across the earth-air interface that accounts for ^{222}Rn and ^{220}Rn in the atmosphere.

6.2 VERTICAL DISTRIBUTION

Once in the atmosphere the radon isotopes continue to decay and as a result the concentration decreases with height above ground. The results of measurements by a number of investigators are shown in Fig. 6.1 The approximate exponential decrease in radon concentration, n, with height above ground, z, is given in eqn. 6.1 which is obtained by the integration of eqn. (6.1). K is the coefficient of vertical diffusion, w is the vertical wind, and λ the decay constant for ^{222}Rn.

$$\frac{\partial n}{\partial t} = \frac{\partial}{\partial z}\left(K \frac{\partial n}{\partial z}\right) - w \frac{\partial n}{\partial z} - \lambda n \qquad (6.1)$$

$$\begin{array}{ccc} \text{turbulent} & \text{vertical} & \text{decay} \\ \text{diffusion} & \text{wind} & \end{array}$$

For steady state and K = constant:

$$n = C_o \exp -\left(\frac{\lambda}{K}\right)^{\frac{1}{2}} z \qquad (6.2)$$

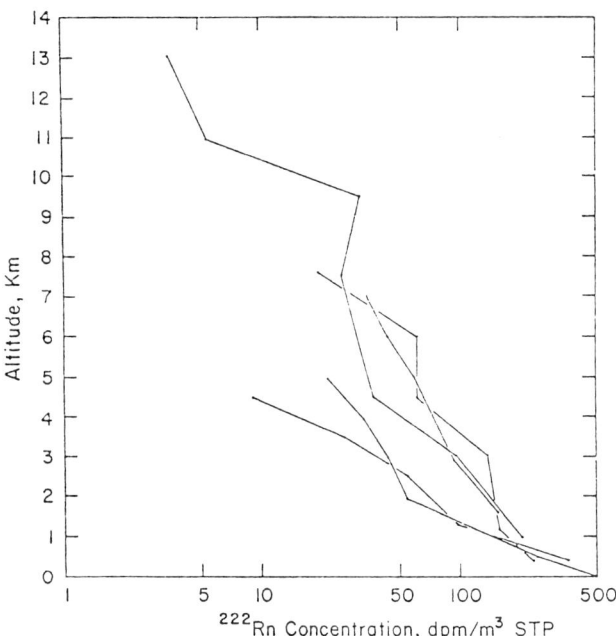

Fig. 6.1 Measured profiles of radon-222 in the lower atmosphere over continental areas. From Moore, et al., 1973, Ref. 6.

A similarity is noted between these equations for transport by turbulent diffusion and vertical wind in the atmosphere and the transport by molecular diffusion and flow below the earth's surface discussed in Chapter 5. However, the effective diffusion coefficients for turbulent processes in the atmosphere exceed those in the soil by a factor of 10^6 to 10^{15}. Hence, it is not surprising that most of the radon in outdoor air originates from the first few meters below the soil surface, while radon can be found throughout the troposphere some 10 to 20 kilometers above sea level. Similarly radon concentrations in the air just above the soil surface can be considered negligible with respect to concentrations in the soil gas only a short distance below. When the wide variation in radon flux density and the vertical mixing processes are taken into account it follows that radon concentrations in the air near ground level

6.3 CONTINENTAL AND MARINE AIR MASSES

A striking contrast exists between typical ^{222}Rn concentrations of about 8 Bq m^{-3} at one meter above ground over continental areas compared with those over oceans at about 0.04 Bq m^{-3}. These figures for ^{222}Rn concentrations are consistent with the large difference in flux density which varies from 17 mBq m^{-2}s^{-1} for soils over the continent to only about 0.1 mBq m^{-2}s^{-1} from the ocean surface (refs. 7, 8).

An increase with distance inland in ^{222}Rn and ^{212}Pb (for ^{220}Rn) concentrations throughout the boundary layer has been measured (ref. 9). The ^{212}Pb (10.5 h) reflects the amount of ^{220}Rn (55 s) released from the land and ocean areas. Radon-222 has been used by Rama (ref. 10) to study monsoon circulation and was able to identify the continental component of monsoon air off the west coast of India. The work of Mishra et al. (ref. 11) using ^{222}Rn and its decay product ^{210}Pb (22 y) and the ^{212}Pb from the thoron (^{220}Rn) series collected at both land and sea stations gives clear evidence of seasonal cycles depending upon the degree of vertical mixing and the type of air masses present.

Collections of air samples at sea have been used to identify radon-rich air masses that have originated from land areas, "radonic storms" in the Antarctic have been studied by Lambert et al. (ref. 12). The French group of scientists has been especially active in their pursuit of air masses that exhibit higher than normal radon activity. Continental air masses over the North Atlantic have been studied based upon both radon and dust content (ref. 13). A good correlation has been shown between ^{222}Rn and dust by a number of observers.

The time scale for transport of air masses over oceans has been determined by Moore et al. (ref. 14) who used appropriate ratios of ^{222}Rn and the ^{210}Pb, ^{210}Bi and ^{210}Po daughters to calculate a transport time of from two to three weeks for air in the trade winds at Hawaii to have crossed the northern Pacific Ocean from the Asian continent. Tracer experiments of this type have good potential for determining air mass trajectories over oceans.

These examples of use of ^{222}Rn in identifying and tracing air masses at sea over continents, and in the study of vertical mixing in the atmosphere demonstrate the usefulness of this isotope in such large-scale studies.

6.4 DIURNAL AND SEASONAL CHANGES

The patterns of daily and annual changes in outdoor ^{222}Rn concentration have been observed in a general way for many years. However, understanding of the physical bases for these changes had to await the development of continuous monitors and a more complete knowledge of transport processes in the atmosphere.

6.4.1 <u>Diurnal variation</u>

Heating of the ground surface by the sun during the day and cooling by radiation during the night causes a marked diurnal change in temperature near the surface. As a result cool air near the ground will accumulate radon isotopes from surface flux during the night, while during the day the warm air will be transported upward carrying radon with it. The result is a significant diurnal fluctuation in radon concentration as illustrated in Fig. 6.2 which gives mean hourly ^{222}Rn decay product alpha activity levels for four months during the year (ref. 15). One hundred alpha counts per minute are equal to 10.7 ± 0.7 Bq m^{-3}. Mean times of sunrise and sunset are given for the months illustrated. The data in Fig. 6.2 were taken as a part of a study covering a 6-year period (692 d of data) in the Rio Grande Valley at Socorro, New Mexico (ref. 16). The results of that study yielded an overall mean of 8.9 ± 2.7 Bq m^{-3}. The month of November has the highest mean at 10.8 Bq m^{-3} while May is the lowest at 7.7 Bq m^{-3}. The average ratio of the daily maximum to the daily minimum was 3.1 with the daily maximum occurring 70 minutes after sunrise and the minimum at 90 minutes before sunset.

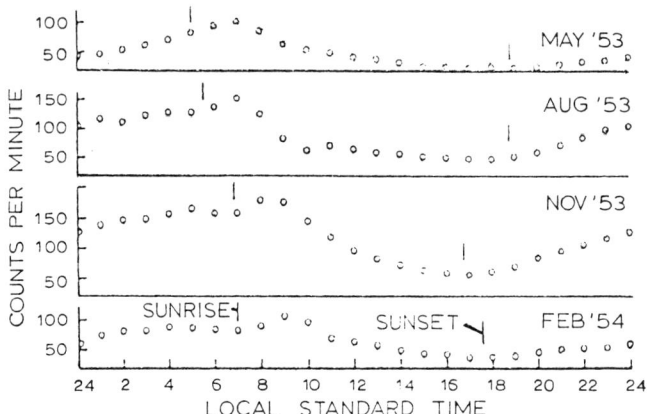

Fig. 6.2 Diurnal course of alpha activity from ^{222}Rn decay products 100 counts per minute equals 10.7 Bq m^{-3}.

Another extensive record of outdoor radon is that carried out by the Environmental Measurements Laboratory in Chester, New Jersey (refs. 17 and 18). A continuous ^{222}Rn monitor has been located at the site some 50 miles from New York City with nine years of data available. A similar monitor has been operated at EML in the city since 1982. Both diurnal and seasonal variations are given in graph form and detailed meteorological information is available. The Chester site yields a mean of 8.1 Bq m^{-3}, a factor of 1.8 higher than average values in the metropolitan area. The mean diurnal ratio of maximum to minimum radon-222 concentrations is 1.8 in Chester compared with 1.3 in Chester.

An interesting treatment of the diurnal pattern for radon concentrations in air near the ground at the Argonne Laboratory near Chicago is given by Moses et al. (ref. 19). Emphasis is given to changes in radon concentrations due to intense thermal convection during the day and strong nocturnal inversions during the night. Detailed plots of temperature, pressure, precipitation, relative humidity, wind and atmospheric stability are given along with the radon concentration.

Additional examples of the variation of radon and/or radon

daughter activities on a diurnal scale are given by Gesell (ref. 20). Included are graphs covering data from New Jersey and New Mexico in the United States, Hungary, Germany, and Moscow and Kirov in the USSR. Two curves are shown for Germany, one for cloudy and the other for sunny days. The night and early morning build up is almost entirely absent for the cloudy days. Annual ratios for diurnal maxima and minima are much less for Germany and Moscow than for the other sites, again reflecting the absence of solar heating during the day on the annual scale.

Typical daily courses of radon concentration are represented in the examples given above. However, exceptions are common when clouds, wind and precipitation have a disturbing effect upon the regular pattern of solar insolation. A rather remarkable exception was a double diurnal maxima in radon concentrations when a monitor was placed on a mountain ridge as shown in Fig. 6.3 (ref. 21). The first maximum occurred when radon accumulated in the air over the mountainous terrain in the usual way in the absence of thermal convection due to solar heating, and the second maximum occurred during the evening after the sun had set but enough heat remained in the mountainside surface rocks to again produce upslope circulation bringing air increasingly enriched by radon in the canyons at the foot of the slope where turbulent convection due to absorbed sunlight no longer occurred. Malakhov et al. (ref. 22) were among the first to show that the diurnal differences in turbulent diffusion and radon exhalation have a direct effect upon disequilibrium among radon and its short-lived daughters.

Analytic treatments of diffusive transport models have been made by a number of investigators.

The earliest, perhaps, was that of Hess and Schmidt in Germany in 1918. A key effort in the application of the theory of turbulent diffusion in the atmosphere to radon concentrations was that of Jacobi and Andre (ref. 23).

Staley (ref. 24) has derived solutions for concentrations of ^{222}Rn, ^{220}Rn and their decay products assuming a sinusoidal diurnal

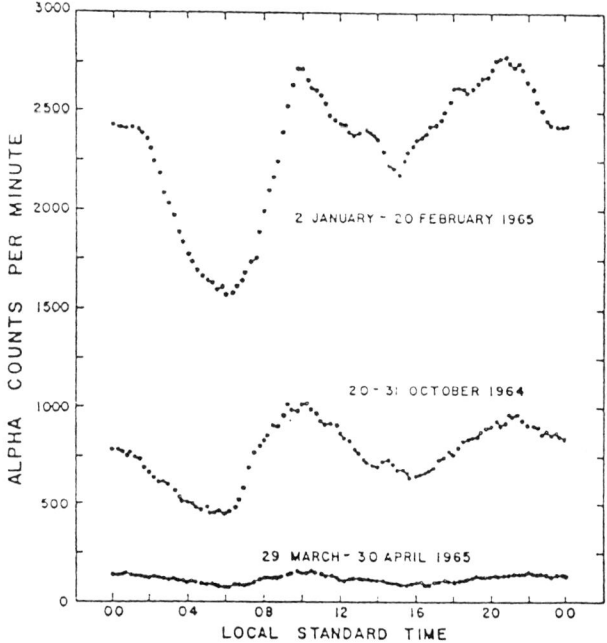

Fig. 6.3 A double diurnal maximum in radon levels on a mountain ridge. (Wilkening and Rust 1972)

time dependence. He found that ^{222}Rn concentrations are influenced mainly by turbulent diffusion in a manner similar to the daily change in potential temperature.

6.4.2 Seasonal variation

Measurements reflecting the mean radon concentrations near ground level at different times of year have been made at a number of laboratories.

Seasonal data from the Environmental Measurement Laboratory at Chester, New Jersey and New York City sites are shown in Fig. 6.4. It is interesting to note that the ratios of the summer maximum to the winter minimum are about the same for the two locations, yet the highest and lowest levels in the city lag those of the Chester area by about one month. No significant correlations between radon

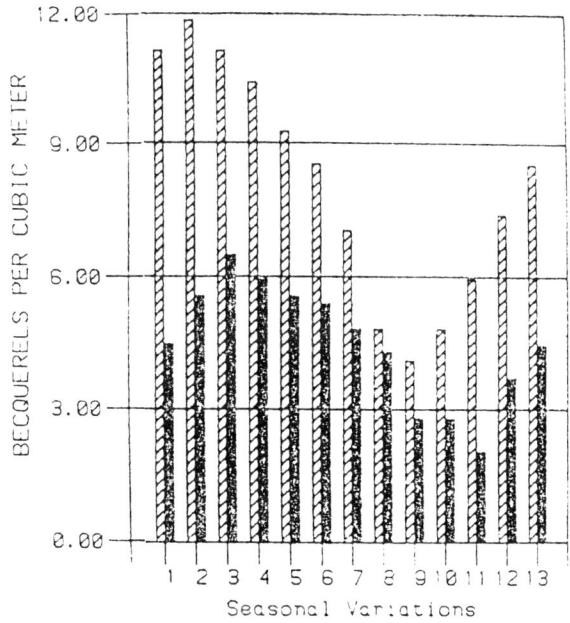

Fig. 6.4 Seasonal variation in radon concentration. Used by permission of the Environmental Measurements Laboratory (ref. 18).
    Nine year (1977 to 1986) average variations in radon concentrations at Chester, NJ (hatched bars).
    Five year (1983 to 1987) average variations in radon concentrations at New York, NY (solid bars).

<u>Seasonal Periods</u>

1	Jul 06-17	6	Nov 23-Dec 04	10	Mar 15-26
2	Aug 03-14	7	Dec 21-Jan 01	11	Apr 12-23
3	Aug 31-Sep 11	8	Jan 18-20	12	May 10-21
4	Sep 28-Oct 89	9	Feb 15-26	13	Jun 07-18
5	Oct 26-Nov 06				

levels and meteorological parameters have been observed.

    The annual course for radon daughter activity from early work in New Mexico (ref. 15) is shown in Fig. 6.5. The data represent monthly means for a four-year period. A limited correlation between natural radioactivity and wind speed can be expected since vertical diffusion is proportional to the mean wind to a limited extent. The spring and early summer months, with high average

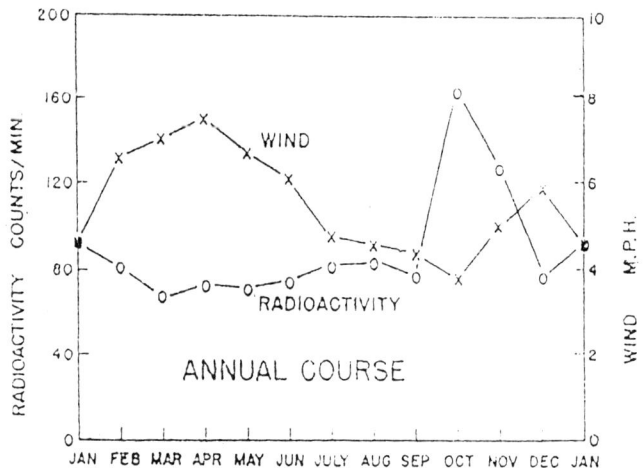

Fig. 6.5 Annual course of radon alpha activity and wind at Socorro, New Mexico. (Wilkening, ref. 15)

wind speeds, show the lowest mean values for natural radioactivity; and the calm, clear months in late fall show predominantly high values of radioactivity. The mean values of natural radioactivity for the months of July, August, and September would undoubtedly be higher if it were not for the summer thunderstorm activity, with the accompanying additional mixing, that occurs during these months in New Mexico. One can be sure that the annual patterns are related to the local climate features. The annual mean radon concentration in the study was 8.9 Bq m^{-3} over a four-year period.

The seasonal variation for the radon at the Argonne National Laboratory in the Chicago area has been reported for a one-year period by Moses, et al. (ref. 19). Highs are evident in the fall months for the early morning, daylight, and evening hours. Moses, et al. attribute the spring minimum to the rainy season at that time of year while pointing out that inversion trapping in the fall months and December contributes to high radon values during these months. The authors note in this paper that "During high winds diffusion rates are high and therefore lower radon concentrations are observed near the ground...." On the basis of these and other measurements it is clear that seasonal variations do occur.

Any analytic treatment of the time and space distribution of radon and its daughters must take into account both the decay characteristics of ^{222}Rn and the meteorological character of the environment. An excellent review of the general theory and a summary of experimental results is given by Reiter (ref. 25).

REFERENCES

1   NCRP 97a, Measurement of radon and radon daughters in air, National Council on Radiation Protection and Measurements, Bethesda, MD, 1988, p. 36.
2   S.D. Schery, D.H. Gaeddert and M.H. Wilkening, Factors affecting exhalation of radon from a gravelly sandy loam, J. Geophys. Res., 89(0-5), (1984) 7299-7309.
3   NCRP 97b, Measurement of radon and radon daughters in air, National Council on Radiation Protection and Measurements, Bethesda, MD, 1988, p. 29.
4   A.B. Tanner, Radon migration in the ground: a supplementary review, T.F. Gesell and W.M. Lowder (Eds.), in: Natural Radiation Environment III, Technical Information Center, United States Department of Energy, 1980, p. 17.
5   C.E. Junge, Air chemistry and radioactivity, Academic Press, New York, NY, 1963, p. 213.
6   H.E. Moore, S.E. Poet. and E.A. Martell, ^{222}Rn, ^{210}Pb, ^{210}Bi and ^{210}Po profiles and aerosol residence times versus altitude, J. Geophys. Res., 78(30), (1973) 7065-7072.
7   M. Wilkening, Radon-222 on the island of Hawaii, Health Phys., 58(1), (1990) 77-80.
8   NCRP 97c, Measurement of radon and radon daughters in air, National Council on Radiation Protection and Measurements, Bethesda, MD, 1988, p. 29.
9   A. Birot, B. Adroguet and J. Fontan, Vertical distribution of radon-222 in the atmosphere and its use for study of exchange in the lower troposphere, J. Geophys. Res., 75(12), (1970) 2373-2383.
10 Rama, Using natural radon for delineating monsoon circulation, J. Geophys. Res., 75, (1970) 2227-2229.
11 U.C. Mishra, C. Rangarajan and C.D. Eapen, Natural radioactivity of the atmosphere over the Indian land mass, inside deep mines, and over adjoining oceans, in: Natural Radiation Environment III, T. Gesell and W. Lowder (Eds.), Technical Information Center, United States Department of Energy, Springfield, VA, 1980, pp. 327-346.

12 G. Lambert, G. Polain and D. Taupin, Existence of periodicity in radon concentrations and in large scale circulation at lower altitudes between 40° and 70° south, J. Geophys. Res., 75, (1970) 2341-2345.
13 R. Larson, and D. Bressan, Radon-222 as an indicator of continental air masses and air mass boundaries over ocean areas, in: Natural Radiation Environment III, T. Gesell and W. Lowder (Eds.), Technical Information Center, United States Department of Energy, Springfield, VA, 1980.
14 H. Moore, S. Poet, E. Martell and M. Wilkening, Origin of ^{222}Rn and its long-lived daughters in air over Hawaii, J. Geophys. Res., 79, (1974) 5019-5024.
15 M.H. Wilkening, Daily and annual courses of natural atmospheric radioactivity, J. Geophys. Res., 64(5), (1959) 521-526.
16 M. Wilkening, Diurnal, seasonal, and geographical variations of outdoor radon concentrations, Figure 7, Health Physics Society Summer School, Las Vegas, NV, 20-25 June, 1982.
17 I.M. Fisenne, Radon-222 measurements at Chester, in: Environmental Measurements Laboratory EML-399, H.L. Volchok (Ed.), United States Department of Energy, New York, NY, 1981, pp. 212-216.
18 I.M. Fisenne and H.W. Keller, Rural and urban radon concentration in Environmental Measurements Laboratory EML-512, Nancy Chieco, Ed.), United States Department of Energy, New York, NY, 1988, pp. 106-108.
19 H. Moses, A. Stehney and H. Lucas, The effect of meteorological variables upon vertical and temporal distributions of atmospheric radon, J. Geophys. Res., 65, (1960) 1223-1238.
20 T.F. Gesell, Background atmospheric ^{222}Rn concentrations outdoors and indoors: a review, Health Phys. 45(2), (1983) 289-302.
21 M.H. Wilkening and D. Rust, Radon-222 concentrations in a mountain canyon environment, Archiv, Meteorol. Geophys. Bioklimatol, A21, (1972) 183-194.
22 S. Malakhov, V. Bakulin, G. Dmitrieva and L. Kirichenko, Diurnal variations of radon and thoron decay products concentrations in the surface layer of the atmosphere and their washout by precipitation, Tellus, 18, (1966) 643-654.
23 W. Jacobi and K. Andre, The vertical distribution of radon-222, radon-220, and their decay products in the atmosphere, J. Geophys. Res., 68, (1963) 3799-3814.
24 D. Staley, The diurnal oscillations of radon and thoron and their decay products, J. Geophys. Res. 75(12), (1966) 3357-3367.
25 E.R. Reiter, Atmospheric transport processes--part 4, Radioactive Tracers, United States Department of Energy, Oak Ridge, TN, 1978, pp. 29-94.

CHAPTER 7

RADON AS A TRACER IN THE ATMOSPHERE

Isotopes of radon found in nature furnish a unique set of tracers for the study of transport and mixing processes in a wide variety of atmospheric phenomena. Radon is a chemically inert substance; and the decay characteristics of its isotopes and their daughter products are precisely known. The ^{222}Rn isotope has been more widely used because of its longer half-life (3.82d) and the greater relative abundance in the free atmosphere; however, thoron (^{220}Rn) because of its short half-life (55 sec) and its origin in the thorium series which is widely distributed in soils and rocks has also found unique applications in tracer studies near ground level.

7.1 TRANSPORT PROCESSES

The physical characteristics of the isotopes of radon and their distribution in the atmosphere where they participate in the fluid character of the atmospheric environment have led to the use of ^{222}Rn, ^{220}Rn and their daughter products as tracers in numerous mixing and transport studies of the atmosphere. The main mechanisms of transport in the atmosphere are horizontal winds, convection and eddy diffusion. Tracer studies have been carried out in (1) the vertical transport of radon by mountain-induced convection, (2) entrainment and mixing in the development of cumulus clouds in New Mexico. Studies of mountain-valley and air drainage have been carried out by a number of investigators.

The relatively high levels of radon encountered in caves have been the target of investigations in the Carlsbad Caverns in the United States, as well as caves in Japan, Hungary, and elsewhere. Tracers such as radon are very effective in identifying natural air exchanges in underground environments. Radon accumulation in caves is directly related to problems found in mines and indoor environments. Natural transport of radon from earth to the atmosphere by volcanoes has also been studied.

The sharp contrast that occurs between ^{222}Rn flux from oceans compared with land areas has led to the identification of substantial differences in radon concentration over those portions of the continents where marine air masses are common. The eastern seaboard of the United States, the Antarctic land mass, and the monsoon patterns in India are areas where such effects are readily identified through radon analyses of the air. Careful studies in these areas have important applications to pollution problems in coastal regions. In contrast with ^{222}Rn, thoron (^{220}Rn) with a half-life of only 55 seconds has been used to study transport processes that occur on a much smaller scale such as those within the first meter or so of the earth's surface.

The short-lived decay products of radon and thoron are isotopes of heavy metals including polonium, lead, bismuth and thallium. Atoms of these elements attach readily to aerosol particles or any solid surface. Characteristics of aerosols have been measured by a careful analysis of the radon decay products which they carry. Such characteristics include size spectra, attachment characteristics, washout and fallout, and residence times in the troposphere, Junge (ref. 1).

The effects of radon and its decay products on the electrical properties of the atmosphere have been described by Israel (ref. 2). The rate at which ions are produced in the air, their numbers, the electrical conductivity and their mobility spectra are all affected by the concentration of radon, thoron and their decay products. The predominantly charged positive ions from the decay of the radon daughters have been used to investigate the atmospheric electrical environments under fair weather and thunderstorm conditions (refs. 3 and 4).

Radon concentrations in soil gas exceed those of the free atmosphere by a factor of about 1,000. As a result there is a flux of radon from soil to air and the transport processes involved in getting the radon atoms from the surfaces of mineral grains to the soil at the air-earth interface form an intricate form of transport in a porous medium (ref. 5). Changes in soil radon concentrations have been used to study stress-strain relations in rock formations and have been proposed for predicting earthquakes. One of the early applications of the use of radon in soil was for uranium exploration.

Radon is readily soluble in water, and its use in studies of the hydrosphere has been underway for some time. The transport of ^{226}Ra and ^{210}Pb (a long-lived daughter of ^{222}Rn) from continents to oceans, the excess of radon in sea water near the ocean floor and the mechanism of radon transport at the air-sea interface are good examples of how radon is used in oceanography (refs. 6 and 7).

The use of radon and its decay products in the study of the earth's atmosphere and hydrosphere has been underway for many years. It can be expected that it will continue in the future.

7.1.1 Radon-222

Radon-222R is especially useful in the study of diurnal mixing patterns and mesoscale systems because of its half-life and the characteristics of its decay products. Early work on time-height relations was carried out by Moses et al. (ref. 8) at the Argonne National Laboratory near Chicago, by Fontan, et al. (ref. 9) near Toulose, and by Cohen et al. (ref. 10) near Philadelphia. These researchers concluded that atmospheric stability as characterized by the potential temperature difference between the ground and a given level in the atmosphere was the most reliable indicator of radon concentration at a given height. According to theoretical treatments ^{222}Rn concentration should decrease with an increase in height in exponential fashion. This effect was confirmed by Moore et al. (ref. 11) using aircraft sampling. A scale height of about 3 km over land areas is typical. Guedalia et al. (ref. 12) have introduced acoustic sounder (sodar) techniques for measuring inversion levels which, when combined with ^{222}Rn concentrations at ground level, allow them to infer variations in mixing heights, vertical diffusion coefficients and the radon flux from the ground. Some other specific examples of the use of ^{222}Rn in atmospheric tracing are given in the following:

Nocturnal air drainage The trapping of atmospheric trace gases and pollutants by inversion at or near ground level is a well-known phenomenon. This is especially noticeable in the case of ^{222}Rn originating from the soil. During the night ^{222}Rn concentration increases when the atmosphere is stable reaching a maximum at near dawn. The effects of mountain-valley air drainage on radon concentrations in the outflow system are less well known. Studies of this type have been made by Wilkening and Rust (ref. 13). Radon-222 concentrations near the mouth of a mountain canyon show diurnal changes consistent with the accepted model of canyon winds wherein surface air moving downslope at night accumulates radon resulting in a steady increase in radon concentration and a decrease in air temperature due to radiation cooling of the slopes above. With the onset of up-canyon winds in the morning, the radon level near the mouth of the canyon drops rapidly to nearly one-tenth of the nocturnal maximum. The steady night-time buildup can be interrupted by wind patterns that intercept the mountain at levels well below the mountaintop or by turbulent mixing induced by lee waves that increase temperature and decrease radon concentrations in the canyon air during the night.

A major effort to delineate flow patterns in mountainous terrain in the vicinity of extensive geothermal development has been carried out in the Geysers area of northern California (ref. 14). In this case point-source releases of tracers including sulfur hexafluoride, perfluorocarbon, heavy methane and fluorescent particles, were used in contrast with ^{222}Rn which is used as a dispersed-source tracer. Results indicate that ^{222}Rn can be used very effectively where tracing from an extended ground-level source is advantageous.

Cumulus convection   Thermal updrafts associated with cumulus cloud development carry ^{222}Rn up in the atmosphere (ref 15). Sampling just below, above and around the perimeter of a developing cumulus congestus cloud demonstrates a positive input of ^{222}Rn into the cloud base and an excess within the cloud in contrast with ambient air outside. The use of ^{222}Rn as a tracer in the study of entrainment, mixing, and outflow patterns during the growth of cumulus cloud systems.

Work along this same line has been reported also by Styro (ref. 16). Radioactivity in cloud droplets as well as in cloud air was measured. Four different patterns of radioactivity versus height were observed. A net flux of radon into the cloud bases was calculated from concentration measurements.

Continental and marine air masses   Measurements have shown that relatively little radon escapes from ocean surfaces compared with the land areas of the world (ref. 17). It can be expected then that radon and its decay product concentrations as well as their diurnal and seasonal variations are quite different in coastal regions compared with inland areas. Guedalia et al. (ref. 18) show that ^{222}Rn and ^{212}Pb concentrations increase with distance inland from the coast throughout the planetary boundary layer. Rama (ref. 19) used ^{222}Rn to study monsoon circulation and determine the continental component of monsoon air off the west coast of India. The work of Mishra et al. (ref. 20) using ^{222}Rn, its short-lived daughters ^{210}Pb and ^{212}Pb at both land and sea stations gives clear evidence of changes in seasonal cycles depending upon the degree of convective mixing and types of air masses present.

Radon-rich air masses over the oceans have been identified that have originated from continental areas. "Radonic storms" were identified by Lambert et al. (ref. 21) as air masses over the Antarctic Ocean that exhibit "pulses" of higher than normal radon activity. Wilkniss et al. (ref. 22) have also measured radon and continental dust in the Antarctic. Larson and Bressan (ref. 23) used ^{222}Rn as an indicator of continental air masses over the North Atlantic, and were able to determine air mass boundaries over ocean areas as well. Prospero and Carlson (ref. 24) found ^{222}Rn and dust coming from the Sahara in a flight over the North Atlantic. A good correlation has been shown between ^{222}Rn and dust by a number of observers.

A study of ^{222}Rn, ^{210}Pb, ^{210}Bi, ^{210}Po and their ratios by Moore et al. (ref. 25) found that from two to three weeks had elapsed since radon in the trade winds at Hawaii had been over the Asian continent. Tracer experiments of this type may have good potential for the identification of air mass trajectories over oceans.

7.1.2 Radon-220 (thoron)

The Israels at Aachen (ref. 26 and 27) were among the first to study ^{220}Rn in air near the ground. An ionization-type instrument was used in their work which was capable of continuously recording ^{220}Rn concentrations at four different heights. At the same time wind speeds and other meteorological data were obtained. A theoretical base was derived including analytical and numerical solutions to the diffusion equation and its boundary value problems. From simultaneous wind and thoron profiles in the interval from 1 to 7.5 m above ground and ^{220}Rn exhalation measurements at the surface, mass and momentum transfer in the first few meters above ground level were obtained. Their measurements also showed effects of wind speed and precipitation on ^{220}Rn exhalation.

Additional work with ^{220}Rn has been done by Crozier and Biles (ref. 28) in the United States; Ikebe and Shimo (ref. 29) in Japan; Druilhet and Fontan (ref. 30) in France; Bakulin et al. (ref. 31) and Filistovich et al. (ref. 32) in the Soviet Union; and Israelsson et al. (ref. 33) in Sweden. Guedalia et al. (ref. 34) measured temperature, wind, ^{212}Pb and ^{222}Rn concentration profiles in the lower atmosphere (100 to 2,000 m) and found reasonable agreement with an "advection" diffusion model. Some studies of thoron have been made more recently by Schery (ref. 35) who has compared the activities of the ^{220}Rn and ^{222}Rn decay products in indoor environments.

Radon-220 has been used primarily in the study of turbulent diffusion and atmospheric electrical parameters within the first few meters of the earth's surface and in examination of meteorological influences on exhalation at ground level. The unique combination of the short-lived ^{222}Rn which is found near the surface and the 10.6 h ^{212}Pb which controls the remainder of the short-lived thoron decay product series provides a challenging opportunity for tracer applications.

REFERENCES

1 C.E. Junge, Air Chemistry and Radioactivity, Academic Press, New York, NY, 1963, pp. 289-348.
2 H. Israel, Atmospheric Electricity Vol. I, Israel Program for Scientific Translations, Jerusalem, 1970, pp. 66-73 and pp. 193-194.
3 J. Bricard and J. Pradel, Electric charge and radioactivity of naturally occurring aerosols, Aerosol Science, (1966) 91-104.
4 A. Roffman, Short-lived daughter ions of radon-222 in relation to some atmospheric processes, J. Geophys. Res., 27, (1972) 5883.
5 M.H. Wilkening, Radon transport mechanisms below the earth's surface, The Natural Radiation Environment III, USDOE and University of Texas School of Public Health April 23-28, Houston, TX 1978.
6 G. Schumann, Radon isotopes and daughters in the atmosphere, Arch. Meteorol. Geophys. Bioklimatol., Ser A, 21, (1972) 149-170.
7 W.S. Broecker, Y.H. Li and J. Cromwell, Radium 226 and radon 222: concentration in Atlantic and Pacific oceans, Science 158, (1967) 1307-1310.

8   H. Moses, A. Stehney and H. Lucas, The effect of meteorological variables upon vertical and temporal distributions of atmospheric radon, J. Geophys. Res., 65, (1960) 1223-1238.
9   J. Fontan, A. Birot, D. Blanc, A. Bouville and A. Druilhet, Measure of the diffusion of radon, thoron and their radioactive daughter products in the lower layers of the earth's atmosphere, Tellus, 18, (1966) 623-632.
10 L. Cohen, S. Barr, R. Krablin and H. Newstein, Steady-state vertical turbulent diffusion of radon, J. Geophys. Res., 77, (1972) 2654-2668.
11 H. Moore, S. Poet and E. Martell, Vertical profiles of ^{222}Rn and its long-lived daughters over the Eastern Pacific, Environ. Sci. and Technol., 11, (1977) 1207-1210.
12 D. Guedalia, A. Ntsila, A. Druilhet and J. Fontan, Monitoring of the atmospheric stability above an urban and suburban site using sodar and radon measurements, J. of Applied Meteorol., 19, (1980) 839-848.
13 M. Wilkening and D. Rust, Radon Transport Processes Below the earth's Surface, Radiation Environment III, T. Gesell and W. Lowder (Eds.), Tech. Info. Center, United States Department of Energy, Springfield, VA, 1980, pp. 90-104.
14 M. Dickerson and P. Gudiksen, ASCOT FY 1979, Progress Report, UCRL-52899, United States Department of Energy, Lawrence Livermore National Laboratory, Livermore, California, 1980.
15 M.H. Wilkening, Radon-222 concentrations in the convective patterns of a mountain environment, J. of Geophys. Res., 75, (1970) 1738.
16 B. Styro and A. Amiranashvili, Some results of the investigations of cumuli clouds natural radioactivity, in: Radioactive Tracers in Research of the Atmosphere and Hydrosphere, Atmospheric Physics 5, MOKSLAS, Vilnius, 1979, pp. 25-42.
17 M.H. Wilkening and W.E. Clements, Radon-222 from the ocean surface, J. Geophys. Res., 80, (1975) 3829-3830.
18 D. Guedalia, C. Allet and J. Fontan, Vertical exchange measurements in the lower troposphere using ThB (^{212}Pb) and radon (^{222}Rn), J. Applied Meteorol., 13, (1974) 27-39.
19 Rama, Using natural radon for delineating monsoon circulation, J. Geophys., Res., 75, (1970) 2227-2229.
20 U.C. Mishra, C. Rangarajan and C.D. Eapen, Natural radioactivity of the atmosphere over the Indian land mass, inside deep mines, and over adjoining oceans, in: Natural Radiation Environment III, T. Gesell and W. Lowder (Eds.), Technical Information Center, United States Department of Energy, Springfield, VA, 1980, pp. 327-346.
21 G. Lambert, G. Polain and D. Taupin, Existence of periodicity in radon concentrations and in large scale circulation at lower altitudes between 40° and 70° south, J. Geophys. Res. 75, (1970) 2341-2345.

22 P. Wilkniss, R. Larson, D. Bressan and J. Steranka, Atmospheric radon and continental dust near the Antarctic and their correlation with air mass trajectories computed from Nimbus 5 satellite photographs, J. Appl. Meteorol., 13, (1973) 512-515.
23 R. Larson and D. Bressan, Radon-222 as an indicator of continental air masses and air mass boundaries over ocean areas, in: Natural Radiation Environment III, T. Gesell and W. Lowder (Eds.), United States Department of Energy, Technical Information Center, Springfield, VA, 1980.
24 J. Prospero and T. Carlson, Radon-222 in the North Atlantic trade winds: its relationship to dust transport from Africa, Science, 167, (1970) 974-977.
25 H. Moore, S. Poet, E. Martell and M. Wilkening, Origin of ^{222}Rn and its long-lived daughters in air over Hawaii, J. Geophys. Res., 79, (1974) 5019-5024.
26 G. Israel, Thoron (Rn-220) measurements in the atmosphere and their application in meteorology, Tellus, 17, (1965) 383-388.
27 H. Israel, M. Horbert and C. de La Riva, Measurements of the thoron concentration of the lower atmosphere in relation to the exchange ('austausch') in this region, Final Technical Report, United States Army Contract DAJA 37-67, C-0593, 1968.
28 W.D. Crozier and N. Biles, Measurements of radon-220 (thoron) in the atmosphere below 50 centimeters, J. Geophys. Res., 71, (1966) 4735-4741.
29 Y. Ikebe and M. Shimo, Estimation of the vertical turbulent diffusivity from thoron profiles, Tellus, 24, (1972) 29-37.
30 A. Druilhet and J. Fontan, Utilisation du thoron pour la determination du coefficient vertical de diffusion turbulente près de sol, Tellus, 25, (1973) 199-212.
31 V. Bakulin, Y. Senko, B. Starikov and V. Trufakin, Investigation of turbulent exchange and washout by natural radioactivity in surface air, J. Geophys. Res., 75, (1970) 3669-3674.
32 V. Filistovich, T. Nedveckaite and B. Styro, On the determination of some turbulent circulation parameters in the near-ground air layer using ^{220}Rn, in Radioactive Tracers in Research of Atmosphere and Hydrosphere, Atmospheric Physics 5, pp. 89-98, MOKSLAS, Vilnius, 1979.
33 S. Israelsson, Meteorological influences on atmospheric radioactivity and its effect on the electrical environment, in Natural Radiation Environment III, pp. 210-225, T. Gesell and W. Lowder (Eds.), United States Department of Energy, Technical Information Center, Springfield, VA, 1980.
34 D. Guedalia, C. Allet, J. Fontan and A. Druilhet, Lead-212 radon and vertical mixing in the lower atmosphere, Tellus, 25, (1973) 381-385.
35 S. Schery, Measurements of airborne ^{212}Pb and ^{220}Rn at varied indoor locations within the United States, Health Phys. 49(6), (1985) 1061-1067.

CHAPTER 8

RADON DECAY PRODUCTS IN THE ATMOSPHERE

Once in the atmosphere the radon atoms continue to decay producing isotopes of polonium, lead, and bismuth. These elements may exist briefly as ions and/or free atoms before forming molecules or attaching to solids in the form of aerosols or other materials. Radon-222 decays with the emission of an alpha particle having an energy of 3.5 MeV. The resulting ^{218}Po atom has a recoil energy of 0.1 MeV sufficient to result in a recoil range of about 50 micrometers in air. The first ionization potential of polonium is only 8.4 eV compared with nitrogen (N_2) at 15.6 eV and O_2 at 12.1 eV, hence the polonium can be expected to be a positively charged ion. The same argument applies to the other decay products at the time of formation. Migdal (ref. 1) concluded in 1941 that the probability for ionization in the outer shells of naturally radioactive atoms at the instant of decay is of the order of unity. They may then form ion complexes or recombine to form neutral atoms either before or after attachment to aerosols where they eventually decay. This has been confirmed by a number of investigators (ref. 2). The characteristics of radon decay products expected in a given air sample will now be addressed in some detail including those present as free ions or atoms and those attached to airborne particles or solid surfaces.

8.1 UNATTACHED RADON DECAY PRODUCTS

At the instant of decay of a ^{222}Rn atom by the emission of an alpha particle the resulting ^{218}Po will be a positively charged ion. Very soon it will pick up an electron to become a neutral atom, but it must be remembered that this ^{218}Po atom has a lower "first ionization potential" than the neighboring O_2 and N_2 molecules; hence it may return to a charged state. The details of the way in which the decay product ions remain positively charged or combine with other atoms to provide "clustering" with polar molecules such as water and sulfur dioxide are not yet fully understood. See refs. 3 and 4. An early analysis of the characteristics of these so-called "unattached" radon decay products was given by Bricard and Pradel (ref. 5) and key reports on their behavior have been made by Porstendorfer (ref. 6) and George and Breslin (ref. 7).

The concentration of the decay product ions in the atmosphere are affected by (1) formation by decay of parent; (2) removal by radioactive decay, by attachment to condensation nuclei, by recombination with negative ions, or by precipitation scavenging, and (3) transport processes including convection, advection, eddy diffusion, sedimentation, and ion migration under the influence of electric fields. A detailed expression of the differential equation for concentration of short-lived ^{222}Rn decay product ions including these terms has been given by Roffman (ref. 8) together with appropriate numerical solutions.

8.1.1 Mean life

The mean life of an unattached ^{222}Rn decay product ion is estimated to fall in the range of 10 to 100 s depending upon (1) its decay constant, (2) the concentration of small negative ions in the atmosphere (n), (3) the density of airborne particles including condensation nuclei (N), and (4) the plateout (PO) on solid surfaces. The mean life, τ, is then given by

$$\tau = (\lambda + \alpha n + \beta N + Po)^{-1} \tag{8.1}$$

where λ is the decay constant for ^{218}Po (3.8×10^{-3} s^{-1}), α is the recombination coefficient for ^{218}Po and ordinary ions of negative charge in the atmosphere, and β is the attachment coefficient. For $n = 0.55 \times 10^9$ negative ions m^{-3} and $N = 10^{10}$ condensation nuclei m^{-3}, the mean life λ is 40 sec if typical outdoor values for the recombination coefficient (1.4×10^{-12} m^3 s^{-1}) and attachment coefficient (2×10^{-12} m^3 s^{-1}) are used. Plateout is considered zero in the outdoor environment where the surface to volume ratio is small compared with the indoor ratio.

A comparison of the magnitudes of the terms in eqn. 8.1 for typical aerosols is instructive. For ^{218}Po, the first radon daughter, which is known to account for approximately 90% of all radon daughter ions, λ is 3.79×10^{-3} s^{-1}. If one uses a recombination coefficient α of 1.4×10^{-12} m^3 s^{-1} appropriate to the positive ^{222}Rn daughter ions and the negative atmospheric small ions (Hoppel, 1969) (ref. 9), a weighted attachment coefficient β of 2×10^{-12} m^3 s^{-1} (ref. 10), a value for the negative total small ion concentration n$^-$ of 0.55×10^9 m^{-3}, and a mean condensation nucleus concentration N of 40×10^9 m^{-3}, one obtains for the "removal" constants for recombination $\alpha n^- = 0.77 \times 10^{-3}$ s^{-1} and for attachment $\beta N = 100 \times 10^{-3}$ s^{-1}. The effective mean life of ^{218}Po ions based upon all three terms given above is 12 s.

It is of further interest to note that the ratio $\beta N / \alpha n = 100$ for the numbers cited indicating that attachment processes dominate over recombination by a large margin.

8.1.2 Mobility of the ^{222}Rn decay product ions

The most important characteristic of an ion apart from its charge and mass is its mobility which is a measure of its drift velocity in an electric field. The mobility of ^{222}Rn decay product ions in the atmosphere has been measured in the range of 0.25 to 1.50×10^{-4} m^2 V^{-1} s^{-1} (ref. 5) which puts them in the same class as ordinary atmospheric small ions resulting from cosmic and terrestrial radiation. These ions in the atmosphere are responsible for such phenomena as air-earth currents and lightning processes.

Measurements have shown that the ^{222}Rn decay product ions make up only about 30 parts per million of the total positively charged ions in clean outdoor air. Simultaneous measurements of the ions from ^{222}Rn and its decay products and the total positive small-ion density in the atmosphere show high correlation coefficients of the order of 0.8 or better for fair-weather conditions in a variety of environments (ref. 3).

8.1.3 Tracers in the atmospheric electrical environment

Studies such as those described above confirm the validity of the use of the radon decay product ions for use as tracers in the study of atmospheric electrical phenomena.

It is quite clear that the positively charged ^{222}Rn decay product ions in the atmosphere are affected by the electric fields of thunderstorms as well as by normal fair-weather electric fields. This is demonstrated conclusively in spite of the fact that decay product ions from radon make up a very minor component of the total positive small ion density.

A study of thunderstorms at the Langmuir Laboratory in New Mexico indicated that both the ions from the radon decay series and the positively charged atmospheric ions were depleted near ground level under the influence of the strong electric fields associated with thunderstorms at the mountain site (ref. 10). These experiments clearly showed that in addition to the major transport mechanisms of eddy diffusion and vertical wind components which apply in both fair-weather and thunderstorm situations, that ion migration is the dominant mechanism for the depletion of the ^{222}Rn decay product positive small ions in the atmosphere near the ground in the presence of electric fields due to thunderstorms. Atmospheric ions including the decay product ions can be removed also by large increases in condensation nuclei to concentrations of the order of 50×10^9 m^{-3} or more, but these conditions are observed less frequently.

8.2 DECAY PRODUCTS ATTACHED TO AEROSOLS

Atoms or molecules including members of the radon decay chains participate in the normal temperature-dependent kinetic behavior of the atmosphere. The high rate of collision among molecules in the air at normal temperatures guarantees that these atoms will collide with airborne particles as well. As a result any system capable of filtering aerosols from the atmosphere whether in ventilation systems or the human bronchi and lungs can be expected to collect radon decay products in varying amounts. It is this mode of exposure that has accounted for the relatively high incidence of lung cancer among uranium miners (refs. 11, 12, & 13), and it has become of increasing concern as a source of exposure in human dwellings.

The major part of the internal dose to humans comes from ^{222}Rn, ^{220}Rn, and their decay products attached to airborne particles. It is important, therefore, to understand the basic features of aerosols in the environment before describing their role in internal dosimetry.

8.2.1 Source and size distributions of aerosols

Aerosols originate from natural and man-made sources in both in both outdoor and indoor environments. Airborne particles in the outdoors are quite different in continental and marine environments. Those found in air above continents stem largely from wind erosion of dry sand and clay soils and from burning and chemical condensation processes of various kinds. Major contributions for the global atmosphere originate in Africa and Asia for the crustal erosion sources and in the industrial areas of the world for the "condensation" type of particles (ref. 14).

The diverse origins of aerosols leads to a wide variation in particle size. In general there are two size groups for both continental and condensation aerosols as shown in Fig. 8.1. A submicron group composed primarily of continental aerosols is centered around a radius of 0.1μm. The smaller particles or "condensation nuclei" result largely from gas-to-particle conversion including condensation of volatile products of combustion, and the direct emissions from the burning of materials.

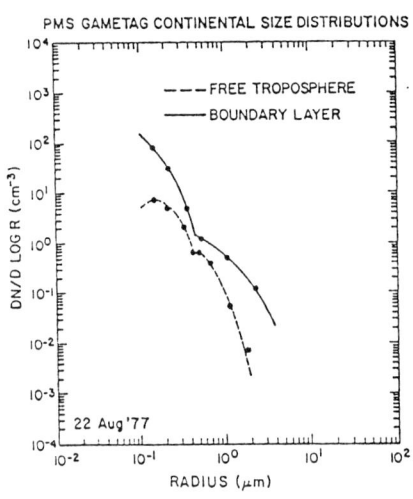 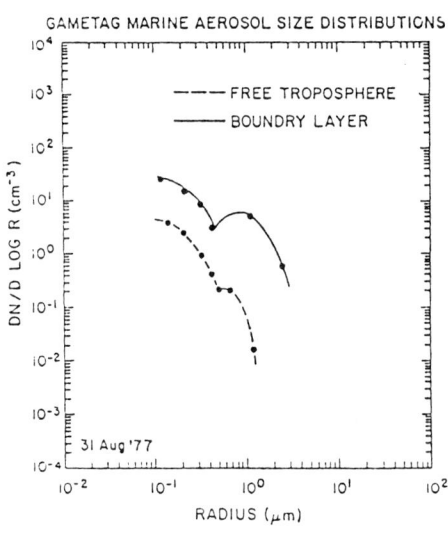

Fig. 8.1. Size-number distributions for continental and marine environments. Solid curves represent data from the boundary layer, 0-2 km (0-6,570 ft) and the dashed curves show size distribution obtained in the midtroposphere at 5.5 km (18,060 ft) MSL. All data obtained aboard aircraft in the GAMETAG program (1977) (ref. 15). Used by permission.

Another common example of this type of particle formation is in automotive exhaust where nitrates are common while sulfates are produced in a variety of human-related activities. Dust from fine grained clay soils contribute to this category also. Small aerosols with radii less than 0.1μ are called Aitken particles. When charged these particles play a role in the electrical conductivity of the atmosphere (ref. 14).

The group with radii larger than 0.5μm contains the crustal aerosols. These particles as well as many of the smaller ones originate in large measure from wind erosion of exposed surfaces. Sandstone based soils contribute to the larger sized aerosols while clays are the major source for the smaller airborne particles. Studies have shown that aerosols originating from the earth's surface have very similar size distributions as the soil particles from which they originated.

Each mode is subject to wide variation in height as shown by the dashed curves of Fig. 8.1. These give size variations of marine aerosols over the Pacific Ocean. The size distribution in the troposphere in each case shows a decrease in density compared with the boundary layer. This is expected in accord with removal of the large particles by sedimentation and the smaller ones by condensation processes. The fact that sedimentation becomes more effective for the larger particles shows up in Fig. 8.1 for a radius greater than about 0.5 μm. The data for continental aerosols are consistent with those for a crustal aerosol.

On the other hand the marine aerosols represented in the boundary layer curve of Fig. 8.1 are formed as salt particles from the sea spray. These particles grow by hydration and then disappear by sedimentation just as the larger particles did in the continental aerosols. In the boundary layers of both continental and marine environments growth mechanisms tend to decrease the number of particles as they grow larger.

In a study of airborne particles in uranium mines George et al. found a log normal distribution with a mean radius of 0.08 μm (ref. 18).

8.2.2 <u>Number concentration</u>

The number of aerosols per unit volume vary widely dependent upon many factors. On the average outdoor and indoor concentrations are roughly the same ranging from 10^3 to 10^5 cm^{-3}. Indoor activities such as smoking, cleaning, and cooking can produce anomalous levels of particle concentrations of the order of 10^6 cm^{-3}. Measurements made in rooms on campus on the New Mexico

Tech campus (ref. 17) gave an average aerosol concentration of 10^4 cm^{-3} with an average deviation of 70%.

8.2.3 Vertical distribution of aerosols

In addition to the variation with size the concentration of aerosols varies with height above ground. For the smaller particles (r < 0.5 µm) the number decreases by about a factor of 10 in going from just above ground level to some 3,500 m above the surface. Large particles (r > 0.5 µm) decrease in concentration by a factor of approximately 150 over the same vertical interval.

The vertical distribution of aerosols both large and small is dependent largely by the amount of convective mixing that occurs. It can be expected therefore, to vary on both diurnal and seasonal scales.

The decay products of radon and thoron in the atmosphere are found in two forms. They may be "unattached" in which case they exist as positive ions or in small molecular clusters in either charged or uncharged states. These make up only about 3% of the total dose from the decay products in a typical environment.

By far the major portion of the radon progeny are found "attached" to airborne particulate matter. Aerosols in the outdoor environment vary widely in size, number concentration, and composition. See ref. 19. One group having a mean diameter of about 0.1 µm is found in concentrations of approximately 3 to 5 X 10^4 cm^{-3} in both indoor and outdoor atmospheres. The larger atmospheric particles have radii of the order of 10-100 µm and originate from crustal erosion over continents or from salt particles in the marine environment. The larger particles will be found in concentrations of approximately one-tenth of that of the smaller particles in outdoor air.

The reader is referred to Deepak (ref. 16) for further information on atmospheric aerosols.

8.3 SUMMARYY

The fact that radon decay products begin their existence as positive ions makes them an interesting and important component of the atmosphere. These ions play important roles both from their contribution to the internal dose to humans and from the effects of ion-pair production on the indoor atmospheric electrical environment.

The main contribution of the radon decay products to internal dose comes from those attached to airborne particulate matter. Aerosols originate from continental areas largely by wind erosion and burning and chemical processes and from salt spray in marine areas.

The concentration of aerosols in indoor and outdoor environments varies widely in the range of 10^9 to 10^{12} m^{-3}. The diverse origins of aerosols leads to two major size groups. One is made up primarily of condensation nuclei about 0.1 μm. Crustal aerosols resulting primarily from wind erosion have radii generally in excess of 0.5 μm. A marked variation of aerosol density with height occurs with figures representing the boundary layer above continents exceeding tropospheric levels by one to two orders of magnitude.

REFERENCES

1   A. Migdal, Ionization of atoms accompanying alpha and beta decay, J. of Physics IV, (1941) 449-453.
2   M. Wilkening, Effect of radon on some electrical properties of air in ACS Symposium Series No. 331, Radon and its decay products, Philip Hopke, (Ed.), American Chemical Society, Washington, DC, 1987, pp. 252-263.
3   V.A. Mohnen, Formation, nature, and mobility of ions of atmospheric importance, in Electrical Processes in Atmospheres H. Dolezalek and R. Reiter (Eds.), Steinkopff, Darmstadt, Federal Republic of Germany, 1977, pp. 1-17.
4   S.N. Rudnick, W.C. Hinds, E.F. Maher and M.W. First, Effect of plateout, air motion and dust removal on radon decay product concentration in a simulated residence, Health Phys., 45, (1983) 463-470.
5   J. Bricard and J. Pradel, Electric charge and radioactivity of naturally occurring aerosols, Aerosol Science, (1966) 91-104.

6   J. Porstendorfer and T.T. Mercer, Influence of electric charge and humidity upon the diffusion coefficient of radon decay products, Health Phys., 37, (1979) 191-199.
7   A.C. George and A.J. Breslin, The distribution of ambient radon and radon daughters in residential buildings in the New Jersey-New York area, in Natural Radiation Environment III, Vol. 2, T. Gesell, W. Lowder (Eds.), Tech. Information Center, United States Dept. of Energy, Springfield, VA, 1980, pp. 1272-1292.
8   A. Roffman, Short-lived daughter ions of radon 222 in relation to some atmospheric processes, J. Geophys. Res., 77, (1972) 30.
9   W.A. Hoppel, Measurement of the mobility distribution of tropospheric ions, Pure and App. Geophys. 81, (1970) 230-245.
10 M.H. Wilkening, Influence of the electric fields of thunderstorms on radon-222 daughter ion concentrations, in Electrical Processes in Atmospheres, H. Dolezalek and R. Reiter (Eds.), Steinkopff Verlag, Darmstadt, 1977.
11 M. Eisenbud, Environmental Radioactivity from Natural, Industrial and Military Sources, 3rd edn., Academic Press, Inc., Orlando, FL, 1987, pp. 173-188.
12 H.T. Miller, Radiation exposures associated with surface mining for uranium, Health Phys., 32(6), (1977) 523-527.
13 J.N. Stannard, Radioactivity and Health, Office of Scientific and Technical Information, Batelle Memorial Inst., Richland, WA, 1988, pp. 113-186.
14 E.M. Patterson, Size distributions, concentrations, and composition of continental and marine aerosols, in: A. Deepak (Ed.), Atmospheric Aerosols, Spectrum Press, VA, 1982, p. 5.
15 E.M. Patterson, C.S. Kiang, A.C. Delany, A.F. Wartburg, A.C.D. Leslie and B.J. Huebert, Global measurements of aerosols in remote continental and marine regions: concentrations, size distributions, and optical properties, J. Geophys. Res., 85, (1980), 7361-7376.
16 A. Deepak (Ed.) Atmospheric Aerosols: Their Formation, Optical Properties, and Effects, Spectrum Press, Hampton, VA, 1982, Chapt. 1.
17 M. Wilkening and E. McNamee, Radon-222 and its short-lived decay products in attached and free-ion forms in the atmosphere, Radiation Protect. Dosim., 24(114) (1988) 221-224.
18 A. George, L. Hinchliffe and R. Sladowski, Size distribution of radon daughter particles in uranium mine atmospheres, HASL-326, Energy Research and Development Administration, New York, NY (1977).
19 H. Pruppacher and J. Klett, Microphysics of Clouds and Precipitation, D, Reidel Pub. Co., 1978, Chap. 8.

CHAPTER 9

EFFECT UPON THE ELECTRICAL CHARACTER OF THE ATMOSPHERE

The presence of the natural radioelements in the atmosphere together with the cosmic radiation and some beta and gamma radiation from the earth's surface results in continuous ionization of some of the nitrogen, oxygen, and trace gas atoms and molecules in the air. The ions resulting from this natural irradiation have a direct effect upon the electrical characteristics of the atmosphere. For example the air is slightly conducting as a result of the omnipresence of these ions. These effects are treated in this chapter.

9.1 IONIZATION

The alpha and beta particles and the gamma rays from ^{222}Rn, ^{220}Rn and their decay products give up their energy by producing ions and raising atoms and molecules to excited states. An alpha particle from the decay of a ^{222}Rn atom (5.49) MeV will produce about 150,000 ion pairs with the expenditure of 34 eV per ion pair (ref. 1). Some details of how these ion pairs are converted into the atmospheric "small ions" were given in Chapter 8.

The chief ionizing agents in air near the ground are the radon and thoron decay chains which yield (4.6 X 10^6 ip m^{-3}s^{-1}. This exceeds that from terrestrial radiation (4.0 X 10^6 ip m^{-3}s^{-1}, and ionization from cosmic rays (1.5 X 10^6 ip $m^{-3}s^{-1}$) (ref. 2).

Ionization rates from radon can vary remarkably ranging from 10 X 10^6 ion pairs $m^{-3}s^{-1}$ in an outdoor mountain environment to 2,300 X 10^6 ion pairs $m^{-3}s^{-1}$ in the Carlsbad Caverns in New Mexico where radon levels are quite high (ref. 3). Measurements of Rn-222 levels in houses in the southwestern United States gave mean Rn-222 concentrations of 63 Bq m^{-3} (ref. 4). An ionization rate of 26 X 10^6 ion pairs $m^{-3}s^{-1}$ can be expected from such concentrations. Allowance for typical Working Levels for Rn-220 (thoron) and its daughters in indoor environments (Schery, 1985) (ref. 5), would increase this figure by perhaps one-third or more. Chalmers in his book, Atmospheric Electricity (ref. 6), gives a figure of 11 X 10^6 ion pairs $m^{-3}s^{-1}$.

Since indoor levels of Rn-222 and Rn-220 and their daughters exceed those of the outdoors by factors of up to 10, it is clear that there is a significant source of ionization in dwellings due primarily to the presence of the radon isotopes and their decay products.

9.2 CONCENTRATION

The concentration of small ions in the atmosphere is determined by 1) the rate of ion-pair production by the cosmic rays and radioactive decay due to natural radioactive substances, 2) recombination with negative ions, 3) attachment to condensation nuclei, 4) precipitation scavenging, and 5) transport processes including convection, advection, eddy diffusion, sedimentation, and migration under the influence of electric fields. A detailed differential equation for the concentration of short-lived Rn-222 daughter ions including these terms as well as those pertaining to the rate of formation of the radioactive positive ions by decay of the parent nuclide and the decay of the ions themselves has been given by Roffman (ref. 7) together with appropriate numerical solutions. Fig. 9.1 taken from Roffman's paper shows a typical diurnal variation of ^{222}Rn decay product positive ions. The effect

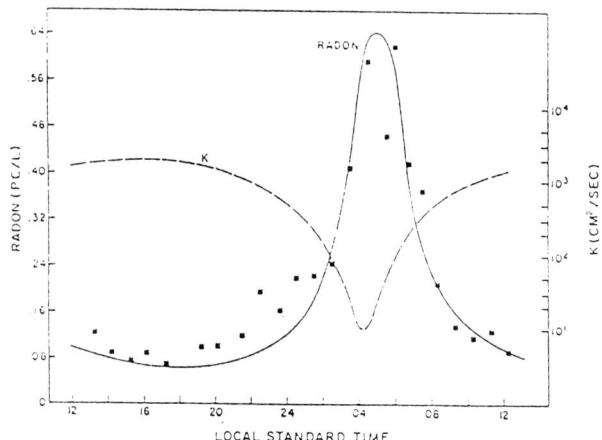

Fig. 9.1 Diurnal variation of the concentration of ^{222}Rn decay product positive ions in the outdoor environment based upon experimental data (black dots and numerical calculations based upon eddy diffusivity (K) at 2 meters above ground shown in the dashed line. (From Roffman, ref. 7).

of normal changes in eddy diffusion (K) is quite clear. It will be noted that the average ^{222}Rn concentration is about 0.1 pCi l^{-1} (3.7 Bq m^{-3}. The early morning peak reaches a level of about 0.35 pCi l^{-1} (13 Bq m^{-3}). The daily pattern for the ^{222}Rn decay product positive ions follows the diurnal variation of the parent ^{222}Rn as expected.

Measurements in a first-floor room on the New Mexico Tech campus (ref. 8) showed an average of 800 ^{222}Rn ions m^{-3} over a five-day period. The total positive ion concentration was 0.1 X 10^9 m^{-3} for the same period giving a ratio of 8 ppm for ^{222}Rn decay product ions to the total small ions of positive charge.

Figures in the literature for the free atmosphere show about 0.5 X 10^9 m^{-3} for both positive and negative ion densities (ref. 9). As expected the total ion density in the free atmosphere exceeds indoor values since plateout is not as great in the outdoor environment.

9.3 CONDUCTIVITY

The electrical conductivity in air is related to the ion density N and the mobility k by $\lambda = Nek$ where e is the electronic charge of the ions. The mobility in this case is the drift velocity acquired in the medium (air) in response to the applied electric field E. For ordinary atmosphere small ions both the N and k values are essentially the same for both positively and negatively charged ions. Typical values are $N = 5 \times 10^9$ ions m^{-3} $k = 1.1 \times 10^{-4}$ m^{2}V^{-1}s^{-1}. The conductivity λ is then in the range of 0.25 to 1.50×10^{-4} m^{2}V^{-1}s^{-1} which is the same expected for ordinary atmospheric small ions (ref. 6). Simultaneous measurements of the concentrations of the Rn-222 daughter ions and the total atmospheric small ions of the same sign yield a correlation coefficient of 0.8 or better for fair weather conditions. This was confirmed by measurements of atmosphere electrical parameters by our group in the Carlsbad Caverns (ref. 3).

The atmosphere is a good insulator but it does have a conductivity which like the ion density is not affected appreciably by the radon daughter ions. It is of interest to note that the electrical conductivity is balanced with respect to positive and negative polarity within the accuracies indicated.

9.4 MOBILITY

A key feature of any atmospheric ion is its electrical mobility which is a measure of its change in drift velocity when placed in an electric field. The mobility depends upon the charge of the ion and inversely upon the mass. The diffusion coefficient, another important characteristic, for an ion is the product of the mobility and its thermal energy. It is mobility that forms the basis for classification of ions. The atmospheric "small ions" have a mobility in the range of 1 to 2×10^{-4} meters/second per volt/meter. They generally consist of a singly ionized atom or molecule with water or other polar molecules clustered about it similar to the radon daughter positive ions described earlier.

The measurements of mobilities of the positive radon decay product ions clearly show that the electrical mobilities and hence, the masses of the radon daughter positive ions fall in the same general range as those for ordinary atmospheric small ions.

9.5 SUMMARY

The decay products of the radon and thoron series are the source of a major part of the ionization in the lower atmosphere exceeding that from cosmic radiation and other sources. A rate of 10×10^6 ion pairs m^{-3} and s^{-1} is typical but subject to wide variation on diurnal and seasonal scales and on geographical location. Ion densities in outdoor air in the lower atmosphere are of the order of 10^8 ions m^{-3} with positive and negative ions occurring in appropriately equal numbers. The ^{222}Rn decay product positive ions are present to the extent of about 800 m^{-3} where outdoor ^{222}Rn levels can be expected to average about 8 Bq m^{-3}. Hence, Rn decay product ions make up only about 8 ppm of the total positive ions in outdoor air. The ions have a mobility in electric fields comparable to those of ordinary atmospheric ions. These properties of the decay product ions have made them useful in tracer studies of a variety of atmospheric phenomena.

REFERENCES

1. H. Israel, Atmospheric Electricity, Vol. II, National Science Foundation, Program for Scientific Translation, United States Dept. of Commerce, Springfield, VA, 1973.
2. CRC Handbook of Chemistry and Physics, 60th edn., 1980, F-212.
3. M.H. Wilkening and V. Romero, ^{222}Rn and atmospheric electrical parameters in the Carlsbad Caverns, J. Geophys. Res., 86 (1981) 9111-9916.
4. M.H. Wilkening and A.W. Wicke, Seasonal variation of indoor radon at a location in the Southwestern United States, Health Phys., 51(4) (1986) 427-436.
5. S.D. Schery, Measurements of airborne ^{212}Pb and ^{220}Rn at varied indoor locations within the United States, Health Phys., 49, (1985) 1061-1067.
6. J.A. Chalmers, Atmospheric Electricity, 2nd. edn., Pergamon Press, New York, NY, 1967, pp. 101-102.

7   A. Roffman, Short-lived daughter ions of radon 222 in relation to some atmospheric processes, J. Geophys. Res., 77, (1972) 5883-5899.
8   M. Wilkening, Effect of radon on some electrical properties of indoor air, in: P.K. Hopke (Ed.), Radon and Its Decay Products, Am. Chem. Society, Washington, DC, 1987, pp. 252-263.
9   H.R. Pruppacher and J.D. Klett, Microphysics of Clouds and Precipitation, Reidel, Holland, 1978, pp. 580-620.

CHAPTER 10

RADON UNDERGROUND

The ^{222}Rn atoms, once formed by the decay of the parent ^{226}Ra, are free to diffuse through the spaces between mineral and soil particles where they become a component of the soil gas. Radon-222 concentrations in the pores and capillaries of soil air a few meters below the surface reach levels of the order 55,000 Bq m^{-3}. Of special interest are ^{222}Rn concentrations in large underground cavities such as caves, tunnels and mines. Elster and Geitel in 1902 noted that mines had high concentrations of the newly discovered "radioactivity." Unless underground cavities have specific natural or artificial ventilation systems, ^{222}Rn levels can be expected to approach the levels found in soil gas.

There are three main cases to be considered. The first requires temperature gradients favorable to vertical convective transport through relatively large openings. The second involves the transport of ^{222}Rn by air movement through cracks, fissures, or underground openings where pressure gradients exist. These occur as the result of changes in barometric pressure, wind-induced pressure or other causes. This will be discussed in relation to the air-exchange characteristics of the Carlsbad Caverns, a large cave complex in New Mexico. The third case is the important one having to do with radon in mines where uranium-radium source

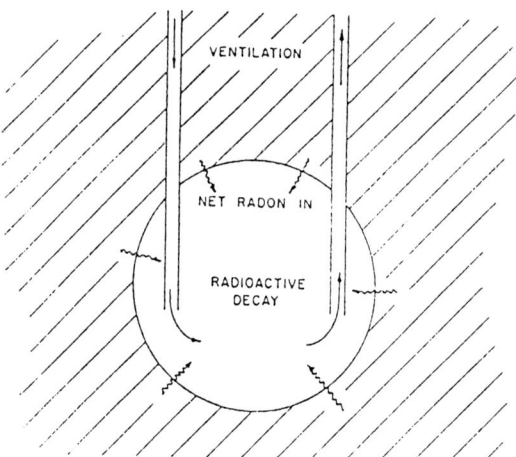

Fig. 10.1 Factors governing the radon balance in an underground cavity (ref. 1).

material is more abundant by a factor of 10^4 or more in such ores.

10.1 RADON IN AN UNDERGROUND CAVITY

Radon concentrations expected in an extended volume within the earth are governed by factors illustrated in Fig. 10.1. The net flux of ^{222}Rn atoms from the walls is balanced by the losses due to radioactive decay and the ventilation process when the system reaches equilibrium. In this case, radon-rich air within the cavity is replaced by air from outside having a low radon content. For a cavity of interior surface S and volume V, this would be

$$SE = \lambda VC + Q(C - C_{od}) \tag{10.1}$$

where E is the net inward flux of atoms per second for each square meter, λ is the decay constant for ^{222}Rn, C and C_{od} are the ^{222}Rn concentrations within the cavity and outdoors respectively in atoms of ^{222}Rn per cubic meter, and Q is the airflow in cubic meters per second.

If one brings air of very low concentration from out-of-doors

($C_{od} \cong 0.2$ pCi/l) relative to that within the cave ($C \cong 40$ pCi/l), C_{od} can be neglected and (10.1) can be solved for the radon concentration giving

$$C = \frac{SE}{\lambda V + Q} \tag{10.2}$$

The radon concentration is seen to depend inversely upon the sum of λV and the air exchange rate Q as well as directly upon the net inward flux, SE. The radon concentration could be calculated for the system shown in Fig. 10.1 if the physical dimensions of the cavity are known and E and Q are measured.

This theory has been applied to the Carlsbad Caverns in New Mexico. In this case the radon-rich air in the Cave is displaced by cooler air from the surface which has a low ^{222}Rn concentration typical of outdoor air. The natural ventilation is assumed to be proportional to the temperature difference between the cave and outdoor air. Almost no exchange occurs in the summer when surface air is warmer than air in the cave. Air exchange does take place in the winter when cooler outdoor air displaces some of the cave air which stays almost constant year around at 14 ± 1°C. The annual variation of ^{222}Rn concentration in the Caverns based upon this model are in good agreement with measured values (ref. 1).

There are many opportunities for this type of air exchange to occur in underground air spaces including those around and within buildings where the driving force may be due to barometric pressure changes, temperature differences or ambient winds.

10.2 RADON IN A TUNNEL

An abandoned tunnel through soil and rocks in a hill west of Socorro, New Mexico is nearly horizontal with a length of 172 m and a cross section of about one by two meters. See Fig. 10.2. The mean depth below the surface is 30 m. The entrance is sealed except for a bypass pipe. No uranium-bearing minerals exist in more than trace amounts.

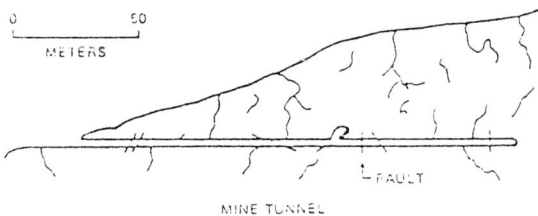

Fig. 10.2 Tunnel in a fractured and geothermally heated hillside. A sealed portal at the tunnel entrance permits measurements of airflow in and out of the tunnel-hillside system. From Natural Radiation Environment III, 1980 (ref. 2).

Air moves into and out of the tunnel due to (1) the diurnal and temporal fluctuation in barometric pressure and (2) convective flow through fissures in the rocks and soil between the tunnel and the outdoor air above the hill. Most of the latter flow is due to a temperature gradient resulting from a geothermal anomaly in the region where the tunnel is located.

The mean temperature inside the tunnel during the measurements was constant at about 18°C. The mean daily outdoor temperature for this same period was 14°C. The mean minimum outdoor temperature was 5°C. A net mean flow into the tunnel was observed which for a 21-day period amounted to 930 m³/day or the equivalent of almost three times the free volume of the tunnel. The mean ^{222}Rn concentration inside the tunnel for the period was 3,300 Bq/m³, with diurnal maxima averaging about 7,400 Bq/m³. The passing of a low-pressure system during the period of study sent ^{222}Rn values up to 19,000 Bq/m³. The latter is near the ^{222}Rn concentration in the cracks and fissures in the rock surrounding the tunnel. The net flow of air into the tunnel each day apparently displaced an equal volume of radon-rich air from cracks and fissures out through the rock and soil above the tunnel. The result is that approximately 220 Bq/sec of ^{222}Rn is added to the air above the hill in excess of the normal transport by diffusion. The tunnel is an example of a natural underground system where flow occurs through crevices and fissures under the influence both of changes in atmospheric pressure and a convective process due to a temperature difference between the tunnel and the surface.

The tunnel in this experiment illustrates how the diurnal atmospheric pressure cycle coupled with the heat source inside the hill can account for the addition of large quantities of radon to the environment. Cavities near the foundation of a house, crawl spaces, or pipe conduits could transfer radon to a dwelling interior in a similar manner. In the case of dwellings transfer of radon-rich air from cavities and channels under and around the foundations may result from barometric pressure changes and flow due to temperature differences. This is an important addition to the radon that comes from building materials, gaseous fuels, water outgassing, and outdoor air.

10.3 RADON IN MINES

It began with the mining of silver in Czechoslovakia in the 15th century. The silver was associated with a black ore called pitchblende from which uranium was identified by the German chemist Klaproth in 1789. One hundred years later at the beginning of the 20th century the Curies isolated radium from pitchblende and the radioactive gas radon was identified almost immediately.

Respiratory disease had been a common illness among the early uranium miners, but was not identified as lung cancer until 1913. It had been called "Bergkrankeit" or mountain sickness. Of those miners who died in those days some 50% had lung cancer. Apart from the usual dangers associated with underground mining, excessive exposure to radon came to be recognized as a major cause of death. Levels of the order of 100,000 Bq m^{-3} (2,900 pCi l^{-1}) were encountered in the Schneeberg (1923) and Jachymov (1924-25) mines. See ref. 3. In a study of the deaths of some 116 miners in Schneeberg and Joachimsthal it was determined that 54% had lung cancer (ref. 3 - Table 3.1).

10.3.1 Uranium mining

When the need for uranium in nuclear programs was recognized during World War II, the United States obtained first some uranium ores from the Big Bear Lake in northern Ontario and substantial quantities of high grade ore from the Belgian Congo.

Uranium fission was discovered by Hahn and Strassmann in 1939. The successful operation of the first nuclear reactor by Enrico Fermi and his collaborators on December 2, 1942 increased the need for uranium in the years ahead in a dramatic way. This unit contained 5,600 kilograms of uranium metal and 36,600 kg (12,400 lb) of uranium oxide in a graphite matrix having an external dimension of some 7.4 m (ref. 4).

The growth of the uranium industry in the United States alone is illustrated by the fact that in December 1975 seven western states had a total of 19 active uranium mills with a nominal capacity of some 20,000 tons of ore per day (ref. 5). Twelve of these mills began operation in the 1950's. By 1977 New Mexico had 5 mills with a capacity of 22,160 tons of ore per day.

Also in 1977 New Mexico had 32 active uranium mines employing some 3,350 miners. The total number of uranium miners in the United States reached a peak of 10,260 in this same year. See ref. 6. The average maximum depth of the New Mexico mines was 238 m (780 ft). Ventilation was accomplished through 107 vents discharging 5 million cfm (140,000 $m^3 s^{-1}$) to the outdoor air.

In the early 1940's uranium mines in Colorado and Utah that had only natural ventilation had levels in the 75,000 to 190,000 Bq m^{-3} (2,000-5,000 pCi l^{-1}) range. Forced ventilation of mines in the late 1930's reduced these levels to about 2,000 Bq m^{-3} (50 pCi l^{-1}). It should be recognized that mines other than those for uranium that have only natural ventilation can reach levels of 40,000 Bq m^{-3} (1,000 pCi l^{-1}).

Another report shows that there were 21,951 employees in the United States uranium industry peak in 1979. This reduced to about 2,140 employees at the end of 1988. By the end of 1989 the last two of some 55 uranium mines in New Mexico were shut down. Homestake closed its uranium mining and milling operations near Grants in November and the Mt. Taylor mine owned by Chevron Resources was closed in December. The price of uranium ore had dropped from a high of $43 per pound in the 1960s to about $9 in the late 1980s.

10.3.2 Radon and aerosols

Attention has already been called to the importance of aerosols as contributors to the dose since most of the radon decay products attach to airborne particulate matter. This is an important characteristic of the mine environment. Some work done by George et al. (1977) (ref. 7) resulted in data on both radon and aerosol concentrations in four mines in New Mexico. The data are shown in Table 10.1.

The mean radon level is 14,200 Bq m^{-3} (380 pCi l^{-1}) which may be considered typical of well ventilated mines. The aerosol concentration varies from mine to mine depending primarily upon the level of mining activity involved, the character of the rock material in which the mine is located, and of course, upon the level of ventilation and filtration that is in operation. The mean density of 13 X 10^{10} particles per cubic meter is about that ordinarily found in underground mines. The activity median diameter (AMD) in the range of 0.09 to 0.3 μm with a mean of 0.18 μm, is a little larger on the average than indoor aerosols (ref. 8, 9).

It is of interest to note that mines B and C where blasting, ore hauling and maintenance activity were underway show aerosol levels from 2 to 3 times that of mines A and D where less mining activity is underway. The aerosol content of air in the mines plays a very substantial role in the total radiation dose along

TABLE 10.1

Grants Uranium Mines. Average radon-222 concentrations and aerosol characteristics are given for four mines. Adapted from: George et al., 1977 (ref. 7).

Mine	Radon (Bq m^{-3})	Particle Size AMD (μm)	Aerosol Concentration (X 10^{10} m^{-3})	Mining Activity
A	7,720	0.20	8.6	None
B	22,600	0.19	16	blasting, slushing, diesels
C	9,300	0.20	22	ore hauling
D	17,000	0.13	5.6	maintenance
MEAN	14,200	0.18 μm	13 X 10^{10} m^{-3}	

with the high radiation levels encountered.

The awareness of a health hazard among miners in mid-Europe began in the 15th and 16th centuries. This illness was associated with silicosis and tuberculosis, but had a definite relation to mining in the area where shortened life spans were quite evident. It wasn't until the 1920's that the "death shafts" of Schneeberg and Joachimsthal were identified as having high levels of radon. The number and quality of measurements of radon progressed through the 1930's until the first serious development of a standard based upon studies of the miners occurred in 1941. Two standards, one each for ^{226}Ra and ^{222}Rn, were finally adopted by the National Bureau of Standards (ref. 9). It stated that a prolonged breathing of air containing ~ 10^{-9} Ci l^{-1} (1,000 pCi l^{-1}) as 37,000 Bq m^{-3} could be associated with an increase in lung cancer among miners. A factor of 100 was introduced because of the uncertainties involved resulting in a figure for acceptable levels of radon of 10 pCi l^{-1}. It should be understood that this figure did not apply to uranium miners per se since uranium mining did not get underway in the United States until after World War II.

TABLE 10.2

Radon released from mines and mills. Adapted from R.H. Johnson Jr., et al., in Natural Radiation Environment (Bombay), Wiley Eastern Limited (1982) pp. 182-183 (ref. 10).

Metal	Type of Mine/Mill	Number of Sources	Radon release per source (Bq y^{-1} x 10^{10})
Uranium	underground	305	25,000
Uranium	open pit	63	7,250
Uranium	mill	20	10,400
Iron	underground	11	560
	open pit	57	70
Copper	underground	15	20
	open pit	46	1,500
Zinc	underground	36	850

10.3.3 Mines other than uranium

Once the hazards due to radon for uranium miners were established it did not take long to recognize that other kinds of mines in hard rock areas had high radon levels as well. This would be expected in retrospect since underground cavities including caves have high levels of radon. All figures for amount of radon released are calculated from measured (or assumed in some cases) radon concentrations, degree of mixing with outdoor air, ventilation rates, and other factors.

It must be emphasized also that the uranium ore production, the ventilation characteristics, and the radon output can be expected to vary widely dependent upon the current status of the mine with respect to its production or close-down status.

The information given in Table 10.2 emphasizes (1) that mines other than uranium mines do indeed supply substantial amounts of ^{222}Rn to the atmosphere and (2) open pit mines as well as conventional underground mines are major contributors to conventional underground mines are major contributors to atmospheric radon. The figures given in the table refer to the uranium industry near its peak in 1977-1978.

As expected the data in Table 10.2 also shows that underground mines exceed open pit mines as sources of radon to the atmosphere. The underground uranium mine supplies radon to the atmosphere at a rate of 25,000 X 10^{10} Bq y^{-1} (6,730 Ci y^{-1}) or an amount equivalent to that supplied by 33,000 km^2 (12,700 sq. mi) of typical New Mexico surface soil having an emanation rate of 0.024 Bq $m^{-2}s^{-1}$ (0.65 pCi $m^{-2}s^{-1}$).

REFERENCES

1 M.H. Wilkening and D.E. Watkins, Air exchange and ^{222}Rn concentrations in the Carlsbad Caverns, Health Phys., 31 (1976) 139-145.
2 M.H. Wilkening, Radon transport processes below the earth's surface, in: T.F. Gesell and W.M. Lowder (Eds.) Natural Radiation Environment III, Vol. 1, Technical Information Center, United States Department of Energy, Springfield, VA, 1980, pp. 90-104.
3 J.N. Stannard, Radon and radon decay products: The saga of the Uranium Miners, in: Radioactivity and Health: A History, Batelle Memorial Institute, Richland, WN, 1988, Chap. 3.
4 E. Fermi, Experimental production of a divergent chain reaction, Am. Jour. of Phys., 20(9) (1952) 536-558.
5 C. Weaver and M. Wilkening, Uranium mining and milling, Environmental effects program plan, Environmental Improvement Division, State of New Mexico, 1978.
6 NCRP Report No. 101, Exposure of the U.S. population from occupational radiation, National Council on Radiation Protection and Measurements, Bethesda, MD, 1989, p. 34.
7 A.C. George, L. Hinchliffe and R. Sladowski, Size distribution of radon daughter particles in uranium mine atmospheres, National Technical Information Service, United States Department of Commerce, Springfield, VA 1977, (HASL-326).
8 NCRP Report No. 97, Measurement of Radon and Radon Daughters in Air, National Council on Radiation Protection and Measurements, Bethesda, MD, 1988, pp. 22-23.
9 NBS Safe handling of radioactive luminous compounds, National Bureau of Standards, handbook, H-27, United States Government Printing Office, Washington, DC, 1941.

10 R.H. Johnson Jr., N.S. Nelson, A.S. Golden and T.F. Gesell, in: Natural Radiation Environment (Bombay), Wiley Eastern Limited, 1982, pp. 182-183.

CHAPTER 11

RADON IN WATER

The water in rivers, lakes and streams can be expected to contain radon in small amounts in view of the solubility of radon in water to the extent of almost 0.25 l kg^{-1} under standard conditions (see Table 3.1) and the wide distribution of uranium and thorium and their decay products in the soil and rocks of the Earth's crust (Table 4.1).

11.1 PUBLIC WATER SUPPLIES

During the condensation and growth of rain drops minute quantities of radon from the air are included such that rain water contains only about 50 Bq m^{-3} (1.4 pCi l^{-1}) and surface waters less than 2,000 Bq m^{-3}. Water from underground aquifers where contact is made with rocks and soils containing normal quantities of uranium and radium will have an appreciable quantity of dissolved radon that may range from 20 to 44,000 Bq m^{-3} (ref. 1).

A study of public ground water supplies that includes 3,318 samples from 42 states shows a geometric mean of 5,180 Bq m^{-3} with a geometric standard deviation of 3,530 Bq m^{-3} (from EPA reports 1979-1983). See ref. 2. In another survey Hess et al. (ref. 3) give a population-weighted average of 6,920 Bq m^{-3} geometric mean for water samples from 2,973 utilities across the United States serving drinking water to some 59 million persons. The lifetime risk per person is 4,400 for the geometric mean with an overall lifetime risk of some 70 parts per million, a very small risk indeed.

It is estimated that some 40% of houses use water from private wells or similar sources having high radon content. Although domestic water use varies widely in the amount of radon released, some uses such as showers and dish washing release up to 98% of the radon in the water to indoor air. Of the 50 to 100 gallons of water used per day per person approximately 70% of the radon is released to the indoor environment. On this basis some 0.5 to 3% of the approximately 40 Bq m^{-3} mean radon concentration of radon in homes originates from the domestic water supply (ref. 1).

From the above studies it can be concluded that radon does not constitute an appreciable hazard in water from public utilities. The interested reader is referred to "Special Issue on Radioactivity in Drinking Water," Health Physics (ref. 4) for further information on this subject.

11.2 SEA WATER

Radium and ^{222}Rn are found in near equilibrium amounts in the oceans at a level of about 4 Bq m^{-3} (0.1 pCi l^{-1}) only about 10^{-3} of that found in domestic water supplies. A deficiency of ^{222}Rn is found near the ocean surface due to escape through bubble formation and evaporation while an excess of ^{222}Rn over its parent ^{226}Ra exists near the ocean floor where minerals in the ocean floor add a small additional component. Since ^{222}Rn atoms cannot escape readily from the water surface compared with the soil and rocks of continental areas, marine air masses contain only about 1% or less radon per unit volume than air over large land areas.

REFERENCES

1. NCRP 103, Control of Radon in Houses, National Council on Radiation Protection and Measurements, Bethesda, MD, 1989, p. 12.
2. W.W. Nazaroff, S.M. Doyle, A.V. Nero and R.G. Sextro, Potable water as a source of airborne ^{222}Rn in U.S. dwellings: A review and assessment, Health Phys., 52(3) (1987) 281-295.
3. C.T. Hess, J. Michel, T.R. Horton, H.M. Prichard and W.A. Coniglio, The occurrence of radioactivity in public water supplies in the United States, Health Phys., 48(5) (1985) 553-586.
4. C.R. Cothern and W.L. Lappenbusch, Guest Editors, Radioactivity in Drinking Water, Health Phys., 48(5) (1985) Pergamon Press, New York, NY, 529-716.

CHAPTER 12

INDOOR RADON

A typical person is said to spend about three-fourths of his/her time indoors. Since outdoor radon concentrations average about 8 Bq m^{-3} compared with indoor levels ranging from 11-300 Bq m^{-3} (ref. 1), it is important to understand some things about the sources and variation of radon in the indoor environment.

12.1 SOURCES AND TRANSPORT

As noted in Chapter 5 the availability of radon in a given place depends upon the "effective" ^{226}Ra concentration, the porosity or degree of channeling within the medium, and a pressure differential or diffusion mechanism suitable to cause a transport of air containing the radon. First, the sources are considered.

Soil. The earth materials immediately below and around a building are the primary sources of radon for the interior. The ^{226}Ra concentrations of soil vary widely from about 10 to 150 Bq kg^{-1} with a mean of approximately 40 Bq kg^{-1}. The "emanating power" which is a measure of the ability of the ^{222}Rn atoms once formed by the decaying ^{226}Ra to escape to the soil air spaces has a typical value of about 0.4. See Chapter 4 for further details. It must be remembered that the building site itself may have been graded or filled in a manner such that the ^{226}Ra content and other soil characteristics are quite different from the original land patterns.

It may be noted in this connection that the diffusion length of radon in soil (the average distance it can diffuse before undergoing radioactive decay) is of the order of 1.6 meters (ref. 2). The comparable figure for concrete is only 0.1 meter. Hence, concrete, bricks and masonry are effective barriers compared with soils. However, when cracks in the construction materials or openings around plumbing develop, the chances for transfer of radon-rich air to the interior is enhanced by a large factor. A clear distinction must be made between diffusion of radon and other gases from the soil through the pores and cracks versus transfer by transport due to pressure differentials. Careful studies of the transfer of radon to houses from soil through basement walls and floors have been made by Nazaroff (ref. 3).

12.1.1 Construction materials

Materials commonly found in building interiors include wood, brick, tile, wall board, concrete, plaster, and insulating materials such as glass and wood. Since these materials have their origins in the earth's crust it can be expected that they will contain natural radionuclides as well. The ^{226}Ra content of some common building materials is given in Table 12.1. It is clear that materials such as concrete, brick and tile contain from 20 to 100 Bq kg^{-1} of ^{226}Ra and while a house may contain several metric tons of masonry and concrete the contribution to indoor radon is relatively small because of the low escape rate of the ^{222}Rn atoms from these solids. While wood has a low ^{226}Ra content (1 Bq kg^{-1}) it is porous and ^{222}Rn can escape in small quantities.

In general, construction materials do not contribute in a major way to the indoor radon concentrations in houses.

12.1.2 Water supplies

Domestic water supplies do contain substantial amounts of

TABLE 12.1

Estimates of ^{226}Ra concentrations in building materials[a] (ref. 4).

Material	^{226}Ra concentration	
	Bq per kilogram	pCi per gram
Wood	1	0.03
Concrete	16-61	0.43-1.65
Brick	42-96	1.1 -2.6
Tile	78	2.1
Wall board		
Natural gypsum	4-10	0.11-0.27
Phosphogypsum	27	0.73
Insulating material		
Glass wool[b]	13-40	0.35-1.1

[a]Adapted from UNSCEAR (1982) page 184.
[b]The impervious nature of these products retards the release of radon.

radon. Of the approximately 5,000 to 7,000 Bq m^{-3} of radon in public water supplies up to 98% is released to the indoor air through activities such as dish washing and showering. Lesser amounts come from other domestic water use (ref. 5). Overall it is estimated that only about 2% of indoor radon comes from the water supply. A careful study of radon from this source in the home has been made in a special issue of Health Physics (ref. 6).

12.1.3 <u>Natural gas and other sources</u>

Natural fuel sources from the earth's crust such as natural gas and coal can be expected to contain radon in appreciable amounts. One study by the United States Environmental Protection Agency showed concentrations of radon in natural gas ranging from about 0.04 to 4 kBq m^{-3} (ref. 7). In general it seems that the contributions of natural gas to indoor radon concentration is not significant.

A summary of information on the sources of radon in houses based upon a literature review is given by Bruno (ref. 8).

12.2 EQUILIBRIUM AND PLATEOUT

The contribution of ^{222}Rn and ^{220}Rn to the radiation dose indoors depends not only upon the concentration of the parent nuclides but also on the degree of equilibrium of the decay products with the parent. The tables of radioactive isotopes A.1, A.2, and A.3 in the appendix show that the short-lived decay products of radon and thoron are isotopes of polonium, lead, and bismuth. These ions/atoms may exist briefly in their ionic or atomic forms, attach readily to aerosol particles or they may plateout on solid surfaces. At any time in this scenario they may be removed by radioactive decay.

12.2.1 Equilibrium fraction

The equilibrium fraction (F) is defined as the degree of radioactive equilibrium between radon and its short-lived decay products.

When ^{222}Rn decays its first four daughter products are isotopes of the heavy metals; polonium, lead, and bismuth with half-lives ^{218}Po (3.1 m), ^{214}Pb (26.8 m), ^{214}Bi (19.7 m), and ^{214}Po (164 μs). Typically this group is in secular equilibrium. If, however, some of the daughters are removed by attachment to solids in the form of dust particles or solid surfaces, or by electrostatic attraction the decay products are in disequilibrium. Typically the daughter products taken together are present at about 50% of equilibrium in indoor environments. The first daughter, ^{218}Po, will in general have a higher equilibrium level (90%) than the next two ^{214}Pb and ^{214}Bi (40% and 30%) as measured in our laboratory at the New Mexico Institute of Mining and Technology. See ref. 9. The values of F vary widely but in general are higher where aerosol concentrations are low as in the outdoors under calm conditions.

A summary of many measurements of the equilibrium fraction in the United States and abroad is given in NCRP Report No. 97 (ref. 10). The mean figure for outdoor values is 0.7 while indoor levels in six countries averaged 0.45. It is important to know whether or not the short-lived daughters are in equilibrium with the radon since the degree of equilibrium figures directly in the dose delivered to living tissues.

12.2.2 Plateout

Reference has been made to "attached" and "unattached" fractions of the radon daughters where airborne particulate matter (aerosols) is the substance to which the daughter ions and/or their molecular clusters attach. A typical indoor environment can be expected to have an aerosol density of some 10^{10} m^{-3} averaging about 0.1 μm in radius (see Sect. 8.2). Ordinary kinetic behavior of the decay product chain allows ready contact with the aerosol particles. The walls, floors, ceilings, and all solid materials within the room also collect radon daughters on their surfaces due to the same kinetic activity. This process is referred to as "plateout." A net flow of radon decay product ions, atoms, molecules, and those attached to very small aerosols can be envisioned as drifting to the solid surfaces.

A drift velocity for these decay products is calculated which is a key factor in the plateout concept wherein the plateout rate is the product of the deposition velocity and the surface area per unit volume. The deposition velocity or drift to a surface has been measured at about 10 m h^{-1} (ref. 11).

12.3 VENTILATION

Since outdoor radon concentrations are usually only of the order of 15% of indoor levels, it can be expected that ventilation effects will be considerable. Even with doors and windows closed the average home will have an air exchange (ventilation) rate of approximately 0.1 to 1 per hour. The rate at which the decay products in ion or cluster form deposit is much higher than the rate for ordinary airborne particles with decay products attached. Hence, clean air with low airborne particle concentrations allows decay products to attach to the walls at a relatively high rate. Therefore, the use of particle cleaning devices tends to lower radon decay product concentrations in the air due to increased deposition on walls and other surfaces (refs. 12 and 13).

It must be recognized that while ventilation usually implies the introduction of outdoor air with low radon levels to replace indoor air, the air intake system must indeed have low levels of radon. Systems that have intakes in basements or crawl spaces may have high radon levels which could increase the indoor hazard rather than alleviating it.

12.4 METHODS FOR CONTROL

An understanding of the sources of radon for a given house is the first step in planning ways for control of levels in the interior. The origin of radon in a typical house is given in Table 12.2. It is clear from this information that the ground under the house is the primary source delivering from one-half to two-thirds of the radon to the building interior. The accumulation of high levels in the basement, crawl spaces, or ground-level floors or slabs can be expected to occur.

12.4.1 Soil

There are two approaches to reducing the transfer of radon from the soil to the house interior. One involves the blocking or sealing of the soil against the escape of radon and the other uses ventilation or air mixing techniques.

Sealing. An examination of ground floors and basements regardless of the construction material will reveal fine cracks or joints which provide routes of entry for radon rich air. Sealant material such as rubber-asphalt mixes can be used. Epoxy and concrete materials are not in general suitable because of their

TABLE 12.2
Approximate contributions from sources of radon in houses[a].

Source	Estimated contribution (activity per second)
Soil gas transport[b]	0-6[c] Bq s^{-1} (0-150 pCi s^{-1})
Release from potable water	0-2 Bq s^{-1} (0-60 pCi s^{-1})
Soil gas diffusion[a]	0.1-0.2 Bq s^{-1} (3-6 pCi s^{-1})
Diffusion from building materials	0.01-1 Bq s^{-1} (0.3-30 pCi s^{-1})

[a]From Bruno (1983)
[b]May be a factor of 10 to 100 times higher in certain regions.
[c]Ranges given are subject to wide variation dependent upon location, water sources, house design and construction materials.

tendencies to crack. So called "membrane sealants" when formed in place have proven especially useful for vertical/horizontal joints such as the juncture of a wall and floor. Openings around sewer and water connections require careful sealing with the same type sealants. In some cases the simplest and most effective solution is to cover the entire base floor and walls with two or more coats of epoxy sealant.

12.5 SURVEYS

Major efforts continue in the United States and abroad to establish reliable figures for radon concentrations in the home and work place. Before meaningful results can be obtained, it must be remembered that wide variation is expected dependent upon both diurnal and seasonal scales in the outdoor environment (ref. 14).

Programs aimed at obtaining reliable data on indoor radon concentrations in countries around the world have become available in only relatively recent times. A concern for natural atmospheric radioactivity came long after an awareness of health hazards encountered by uranium miners, luminous dial painters, users of X-rays in medicine, and scientists involved in the development of nuclear energy.

TABLE 12.3

Surveys of indoor radon.

Source	Location	Number Houses	Radon (Bq m^{-3})	Comment
B.L. Cohen(A) Univ. of Pittsburgh	U.S.	50,000	65	49 states range 28-131 Bq m^{-3}
A. Nero, et al.(B) Lawrence Berkeley Lab.	U.S.	817	55	21 states range 17-282 Bq m^{-3}
W. Nazaroff and A. Nero, Lawrence Berkeley Lab.(C)	Canada	9,999	13*	
	Sweden	500	103	
	Denmark	22	70	
	Finland	2,000	64*	
	F.R. Germany	6,000	49	
	Netherlands	1,000	24*	
	Belgium	79	41*	
	France	765	76	
	Ireland	250	43*	
	Japan	251	19*	
B.M.R. Green, et al.(D)	United Kingdom	2,093	22	standard dev. 46 Bq m^{-3}
Pan Ziqiang et al.(E)	Beijing China	364	34.5	range 2-260 Bq m^{-3}
U.C. Mishra and C. Subba Ramu(F)	Bombay India	100	19.0	1.5 outdoors F = 0.43

References
 A NCRP, Proc. No. 10, 1989, p. 179
 B Science, 21 Nov. 1986, p. 992
 C Radon in Indoor Air, Nazaroff and Nero, p. 12
 D Radiat. Prot. Dosim. 24(1-4) 1988, p. 541
 E Radiat. Prot. Dosim. 24(1-4) 1988, p. 32
 F Radiat. Prot. Dosim. 24(1-4) 1988, p. 27

*Geometric means, all others are arithmetic.

12.5.1 International

The data shown in Table 12.3 give results of some indoor radon measurements in the United States, China, and India. Differences in annual mean temperatures and life styles may account in part for the lower indoor radon values in the Asian areas. The same study including nine major cities in India showed a mean radon level of 138 Bq m^{-3}.

12.5.2 United States

The Cohen data given in Table 12.3 were obtained from diffusion barrier charcoal adsorption collectors mailed to homeowners who received a modest compensation for their return (ref. 15). Information with respect to floor level, room type, region within the United States, season, age and cost of house, and weatherization were obtained also.

An earlier study by Cohen (ref. 16) gave the result of radon measurement in the homes of 453 physics professors at 101 universities and colleges in 42 states. This survey yielded an arithmetic mean of 54 Bq m^{-3} - almost identical to the results of the Nero, et al. study given in Table 12.3.

The results of an extensive set of measurements of radon-222 in United States homes is given by A. Nero, et al. (ref. 17). Concentrations were measured in 1,381 houses in 21 states from California to Maine. Since some of the data sets favored locations where radon concentrations are known to be high a later analysis (ref. 12b) provided the data shown in Table 12.3 for A. Nero, et al.(B). The arithmetic mean for the 38 locations was 55 Bq m^{-3} (8 pCi/l). This same data set indicated that from 1 to 3% of the single-family houses in this group exceeded 300 Bq m^{-3} (8 pCi/l). A variety of measurement techniques were used including track-etch detectors, continuous radon monitors, activated charcoal, and grab sampling.

12.5.3 Radon measurement at the state level

In cooperation with a program organized by the United States Environment Protection Agency, a number of states began indoor radon surveys in the late 1980's. These efforts had been brought on in part by the discovery of unusually high indoor radiation levels in the Reading Prong area of New Jersey and Pennsylvania.

<u>New Mexico</u>. A statewide radon survey was conducted in New Mexico during 1989 in conjunction with the United States Environmental Protection Agency program. Charcoal canisters were distributed to randomly selected homes in all 33 counties. Preliminary results showed that 75% of the houses involved had levels less than 150 Bq m^{-3} (4 pCi/l), 24% were between 150 and 740 Bq m^{-3} (4 and 20 pCi/l), and only 1% had levels greater than 750 Bq m^{-3} (20 pCi/l).

12.6 SUMMARY

Concern in recent years about environmental radioactivity has led to numerous extensive surveys of indoor radon in houses around the world. The results are not yet definitive in many respects except to emphasize an overall range of from 20-100 Bq m^{-3} with a high variability including one India study that showed some 9 cities with a mean of 138 Bq m^{-3}. The mean for the 74,270 houses given in Table 12.3 yields an unweighted mean of 45 Bq m^{-3}. In the following chapter the health effects of exposure to indoor radon concentrations at the higher levels will be undertaken.

Typical outdoor air at about 8 Bq m^{-3} in continental air and (0.4 Bq m^{-3}) in marine environments such as Hawaii are low indeed compared with the indoor environment.

The chief source of radon in indoor air is from the soil under and around the basement and/or foundation. Soil gas transport to the interior far exceeds the combined contribution from building materials, drinking water, natural gas, and other sources.

Control of radon entry in houses can be accomplished by sealing soil surfaces and cracks and openings around plumbing. Fan systems that provide mixing and create positive pressure differentials in critical areas by use of outdoor air low in radon can be very effective also.

Ventilation properly arranged to minimize problems with heating and cooling systems can be very effective and can be accomplished at relatively low cost if properly planned.

REFERENCES

1. NCRP 97, Measurement of Radon and Radon Daughters in Air, National Council on Radiation Protection and Measurements, Bethesda, MD, 1988, p. 30.
2. A.B. Tanner, Radon migration in the ground: A supplementary review, in: T.F. Gesell and W.M. Lowder (Eds.) Natural Radiation Environment III, National Technical Information Service, United States Department of Commerce, Springfield, VA, 1980, p. 5.
3. W.W. Nazaroff, Predicting the rate of ^{222}Rn entry from soil into the basement of a dwelling due to pressure-driven air flow, Radiation Protect. Dosim., 24(1-4) (1988) 199-202. (Proc. of NREIV, Lisbon, Portugal, Dec. 1987.)
4. NCRP 103, Control of Radon in Houses, National Council on Radiation Protection and Measurements, Bethesda, MD, 1979, p. 11.
5. NCRP 103, Control of Radon in Houses, National Council on Radiation Protection and Measurements, Bethesda, MD, 1979, p. 13.
6. Health Physics, Radioactivity in drinking water, Health Phys. 48(5) (1985) 529-699.
7. EPA Environmental Protection Agency, Assessment of Potential Radiological Effects from Radon in Natural Gas, EPA 52011-73-004, National Technical Information Service, Springfield, VA, 1973.
8. R.C. Bruno, Sources of indoor radon in houses: a review, J. Air Pollut. Contr. Assoc., 33 (1983) 105.
9. M.H. Wilkening, Effect of radon on some electrical properties of indoor air, Radon and its Decay Products, ACS Symposium Series 331, American Chemical Society, Washington, DC, 1987, 252-263.
10. NCRP 97, Measurement of Radon and Radon Daughters in Air, National Council on Radiation Protection and Measurements, Bethesda, MD, 1988, p. 24.
11. J. Porstendorfer, Behavior of radon daughter products in indoor air, Radiation Prot. Dosim. 7 (1984) 107.
12. A. Nero, Earth, air, radon and home, Physics Today, April (1989) 34-a and 35-b.
13. NCRP 103, Control of Radon in Houses, National Council on Radiation Protection and Measurements, Bethesda, MD, 1989, pp. 33-38.
14. M. Wilkening and A. Wicke, Seasonal variation of indoor ^{222}Rn at a location in the southwestern United States, Health Phys. 54(4) (1986) 427-436.
15. B.L. Cohen, Measured radon levels in U.S. homes, Proceedings No. 10, 24th Annual Meeting of the National Council on Radiation Protection and Measurements, National Academy of Sciences, March 1988, Washington, DC, 170.
16. B.L. Cohen, A national survey of ^{222}Rn in U.S. homes and correlating factors, Health Phys. 51(2) (1986) 175-183.

17　A. Nero, M. Schwehr, W. Nazaroff and K. Revson, Distribution of airborne radon-222 concentrations in U.S. homes, Science 234 (1986) 992-997.

CHAPTER 13

HEALTH EFFECTS

The human environment is one in which ionizing radiations are present at all times and at all places on the earth from the deepest caves to the highest mountains and on into space. Radon and its decay products are present wherever radium and thorium exist in the earth or in any planetary material. In the terrestrial environment the inhaled radon isotopes and their daughter products make up almost two-thirds of the total dose to living tissue. Attention is directed now to a more detailed description of the radiations from natural radionuclides and their effects upon living tissue.

13.1 RADIATION FROM RADON AND ITS DECAY PRODUCTS

First the characteristics of the radiations. The decay schemes for ^{222}Rn and ^{220}Rn are shown in Tables A.1 and A.2 in the Appendix where the alpha, beta and gamma radiations are identified together with the principal energies and the mode of decay for a given nuclide in the series. These are the radiations associated with the natural decay series in the air.

The nature of the interaction of these radiations with matter are quite different.

13.1.1 Quantities and units

Before examining the effects of the radiations from radon and its decay products certain definitions regarding exposure and dose are given in Table 13.1.

Some definitions of quantities in the table are in order. The Potential Alpha Energy Concentration (PAEC) is the concentration of radon daughter products, in air, in terms of the alpha energy that will be released during complete decay through ^{214}Po.

The Equilibrium Equivalent Concentration (EEC) is calculated from the radon concentration in equilibrium with its short-lived daughters that has the same potential alpha energy per unit volume as exists in a sample mixture. Basically it amounts to 5.57 X 10^{-9} J/Bq for ^{222}Rn and 7.56 X 10^{-8} J/Bq for ^{220}Rn.

The Working Level (WL) is any combination of short-lived radon daughter products in one liter of air that will result in the emission of 1.3 X 10^5 MeV of potential alpha energy.

The information in Table 13.1 should be of help in making the transition from radon activity and concentration to exposure and dose. Characteristic energies for the alpha particles from ^{222}Rn and its short-lived decay products is 6.17 MeV (^{222}Rn, ^{218}Po and ^{214}Po). The corresponding range in air for an alpha particle of this energy is about 5 cm, in living tissue it would be only about 0.05 mm. The beta particles (electrons) have a characteristic end point (maximum) energy of about 1.1 MeV. Those with this energy could penetrate about 1.5 mm in tissue. The beta particle of average (not maximum) energy would have a much shorter range. The gamma rays from ^{212}Bi with a mean energy of about 1 MeV would not be absorbed to any appreciable extent in the local tissue.

It is the alpha particles that can be expected to expend their energy in the sensitive surface tissue in the bronchial epithelium or lung. The beta and gamma rays play only a very minor role.

TABLE 13.1

Radiation quantities and units. From Nazaroff and Nero, Eds. Radon in Indoor Air, 1988, p. 39.

Quantity	SI Units	Equivalents
Activity (Becquerel)	Bq	1 curie = 3.7 X 10^{10} Bq 1 pCi = 0.037 Bq
Concentration	Bq m^{-3}	1 pCi l^{-1} = 37 Bq l^{-1}
Equilibrium Equivalent Concentration	EEC_{222} EEC_{220}	1 WL = 3,740 Bq m^{-3} 1 WL = 276 Bq m^{-3}
Absorbed dose (Gray)	Gy	1 Gy = 100 rad
Dose equivalent (Sievert)	Sv	1 J kg^{-1} = 100 rem
Working Level	WL	1 WL = 1.3 X 10^5 MeV l^{-1}
Working Level Month	WLM	1 WLM = WL $\frac{hours}{170}$
Potential Alpha Energy Concentration (no longer in use).	PAEC	1 PAEC = 1 J m^{-3}

13.2 THE HUMAN DOSE

The radon and its decay products in both "attached" and "unattached" forms are delivered to sensitive tissue in the human respiratory system. An adult not engaged in more than light activity can be expected to breathe at a rate of about 0.75 m^3 min^{-1}. Typical aerosol concentrations indoors or outdoors are of the order of 10^{10} per cubic meter with radii centered around 0.5 X 10^{-6} m. As indicated in Chapter 8 indoor air can be expected to contain some 50 Bq m^{-3} of radon. Hence, some 40 Bq of radon are taken into the lungs per minute with perhaps about 3% of the daughter-product dose on the unattached fraction made up of ions and molecular clusters of very small size.

Once in the upper respiratory system the aerosol particles with the radon decay products attached find their way to the bronchial tissue in the lungs. Typically the surface walls of the bronchii leading to the lungs is where most damage by radiation from the decay product alpha particles occur. The mucous layer on the surface of the bronchial tubes is not heavy enough to absorb alpha particles which can damage the basal cells underneath.

It is clear that lung dose models must take into account a complex combination of factors including the activities of the short-lived daughters of ^{222}Rn as a fraction of their equilibrium activities in the decay chain, the half-lives of these daughters, deposition probabilities for attached and unattached activities, fraction of the activities deposited in the tracheo-bronchial region, and allowance for natural clearance of aerosols from the surface tissues.

Progress is substantial in the direction of the development of a model for calculating risk from exposure to radon. A recent lecture by Bo Lindell gives an excellent example of progress in this area (ref. 1)

Much attention has been given to the hazards to uranium miners with regard to exposure to radon and its decay products. Studies are being continued on miners in New Mexico in a recent paper as indicated by Samet (ref. 2).

13.3 IMPACT

The compelling effects of radon and its short-lived decay products spread slowly but surely through a wide range of biological problems encountered in such areas as the mortality rates and lung cancer in uranium mines, the results of experimental work with animals, and the discovery of unusually high levels of radon in the living environments of the general population. The increased concern on the part of scientists and governments around the world has been reflected in international meetings such as those sponsored by the Commission of the European Communities, the United States Department of Energy and other agencies beginning with the first one on the Natural Radiation Environment in Houston in 1963, and followed by other meetings in 1972 and 1978.

Additional ones were held in Brazil (1975), Bombay (1981), Anacapri (1983), Maastricht (1985), and Lisbon (1987). Another in the series, Natural Radiation Environment V, is planned for 1991.

A number of books on radon have been published including recent ones on indoor radon edited by Nazaroff and Nero (1988) (ref. 3) and Bodansky et al. (1987) (ref. 4). The Proceedings of the 24th Annual Meeting of the National Council on Radiation Protection and Measurements entitled Radon (1989) is edited by N. Harley and gives much information on recent studies (ref. 5). Programs set up in a number of states including those in New Jersey, Pennsylvania, and Florida are described. Federal programs in the United States directed by the Environmental Protection Agency, the Department of Energy, and the Mine Safety and Health Administration are included with an outline of their efforts.

13.4 SPAS

The radon chronicle would not be complete without mention of the degree of exposure in the spas and therapeutic galleries mentioned in Section 10.4. It is not unusual for hot springs and other natural waters to contain high concentrations of radon.

Most famous of the European spas and galleries are those such as Badgastein situated in a narrow valley of the Austrian Alps. This spa has been in operation for some 600 years. There are nine thermal springs near the center of the town providing some 5,000 m^3 of water per day at a concentration of some 1.5×10^9 pCi/m^3. Water from the wells is collected in large reservoirs and from there delivered to some 120 hotels and treatment centers. Much of the radon escapes from the water to the atmosphere in transit and while being used in bath tubs and treatment rooms. A former gold mine is the "Thermal Gallery" where temperatures reach 41°C and relative humidity is 99%. The radon level here is 1.1×10^5 Bq m^{-3} (3,000 pCi/l). The ^{222}Rn decay products have an equilibrium factor of about .75 giving a dose of some (22 WL) with an Equilibrium Equivalent Concentration (EEC) of 82,000 Bq m^{-3}.

Another famous radon spa in the area is the one at Bad Hofgastein. There are no hot springs in the area so approximately 1,000 m^3 of water with a radon content of 1.5 Bq m^{-3} (0.04 pCi/l) is supplied through an 8 km long pipeline.

The role of the spas and therapeutic galleries in the history of radioactivity and health is remarkable indeed (ref. 6). From the spas and therapeutic galleries where healing was emphasized, to the uranium mines where lung cancer was a problem, to the mountains of New Mexico where it was used as a tracer in developing storm systems, radon has played important and unique roles.

REFERENCES

1. B. Lindell, How safe is safe enough?, in: N. Harley (Ed.), Proceedings of the 24th Annual Meeting of the National Council on Radiation Protection and Measurements, Proceeding No. 10, NCRP, 7910 Woodmont Avenue, Bethesda, MD, 1989, pp. 196-241
2. J. Samet, D. Pathak, M. Morgan, M. Marburg, C. Key and A. Valdiva, Radon progeny exposure and lung cancer risk in New Mexico uranium miners: a case-control study, Health Phys. 56(4) (1989) 415-421.
3. W. Nazaroff and A. Nero (Eds.), Radon in Indoor Air, John Wiley and Sons, New York, 1988.
4. D. Bodansky, M. Robkin and D. Stadler (Eds.), Indoor Radon and its Hazards, Univ. of Washington Press, Seattle, WA, 1987, pp. 76-90.
5. N. Harley (Ed.), Proceeding No. 10, Twenty-fourth Annual Meeting of the National Council on Radiation Protection and Measurements, NCRP, 7910 Woodmont Avenue, Bethesda, MD, 1989.
6. J. Pohl-Rüling, F. Steinhausler and E. Pohl, Radiation exposure and resulting risk due to residence and employment in a radon spa, in: K.G. Vohra, et al. (Eds.), Natural Radiation Environment, Wiley Eastern Limited, Bombay, 1982, 107-113.

APPENDIX

TABLE A.1

The Uranium Series (4n+2) - From Table of Radioactive Isotopes (ref. 2).

Historical Designation	Element	Symbol	Radiation Energy (MeV)	Half-Life
Uranium I ↓	Uranium	^{238}U	α 4.15	4.47×10^9 y
Uranium X_1 ↓	Thorium	^{234}Th	β 0.20, 0.10	24.1 d
Uranium X_2 ↓	Protactinium	^{234}Pa	β 2.3	1.17 m
Uranium II ↓	Uranium	^{234}U	α 4.72, 4.80	2.48×10^5 y
Ionium ↓	Thorium	^{230}Th	α 4.62, 4.69	7.54×10^4 y
Radium ↓	Radium	^{226}Ra	α 4.78	$1,600 \times 10^3$ y
Ra Emanation ↓	Radon	^{222}Rn	α 5.49	3.82 d
Radium A 99.98% | 0.02%	Polonium*	^{218}Po	α 6.00	3.11 m
Radium B ↓ — Astatine-218	Lead	^{214}Pb	β 0.67, 1.02 0.72	26.8 m
Radium C 99.96% | 0.04%	Bismuth**	^{214}Bi	β 1.54, 3.27	19.9 m
Radium C' ↓ — Radium C"	Polonium	^{214}Po	α 7.69	164 μs
Radium D ↓	Lead	^{210}Pb	β 0.02, 0.06	22.3 y
Radium E ~100% | 2×10^{-4}%	Bismuth***	^{210}Bi	β 1.16	5.01 d
Radium F ↓ — Thallium-206	Polonium	^{210}Po	α 5.30	138 d
Radium G (End Product)	Lead	^{206}Pb	Stable	

* Uranium X_2 is an excited state of ^{234}Pa and undergoes isomeric transition (§ 10.147) to a small extent to form uranium (^{234}Pa in its ground state); the latter has a half-life of 6.7 h, emitting beta radiation and forming uranium II (^{234}U).

TABLE A.2

The Thorium Series (4n)--From Table of Radioactive Isotopes (ref. 2).

Historical Designation	Element	Symbol	Radiation and Energy (MeV)	Half-Life
Thorium	Thorium	^{232}Th	α 4.0	1.41×10^{10} y
Mesothorium I	Radium	^{228}Ra	β 0.04, 0.02	5.75 y
Mesothorium II	Actinium	^{228}Ac	β 2.07, 1.73	6.13 h
Radiothorium	Thorium	^{228}Th	α 5.34, 5.42	1.91 y
Thorium X	Radium	^{224}Ra	α 5.69, 5,45	3.66 d
Th Emanation	Radon	^{220}Rn	α 6.29	55.6 s
Thorium A	Polonium	^{216}Po	α 6.78	0.15 s
Thorium B	Lead	^{212}Pb	β 0.33, 0.57	10.6 h
Thorium C 64% | 36%	Bismuth	^{212}Bi	β and α 2.25, 0.55	60.5 m
Thorium C'	Polonium	^{212}Po	α 8.78	0.29 μs
Thorium C"	Thallium	^{208}Tl	β 1.79, 1.28 1.52	3.06 m
Thorium D (End Product)	Lead	^{208}Pb	Stable	-

TABLE A.3

The Actinium Series (4n + 3)--After NCRP No. 97 (ref. 3).

Historical Designation	Element	Symbol	Radiation	Half-Life
Actinouranium I ↓	Uranium	^{235}U	α 4.40	7.13 X 10^8 y
Uranium Y ↓	Thorium	^{231}Th	β 0.09, 0.30	25.5 h
Protactinium ↓	Protactinium	^{281}Pa	α 5.0	3.2 X 10^4 y
Actinium 98.8% | 1.2% ↓	Actinium	^{227}Ac	α 0.05	21.6 y
Radioactinium | v Actinium X	Thorium Radium	^{222}Th ^{223}Ra	α 5.8, 6.0 α 5.5, 5.7	18.2 d 11.4 d
Actinon ↓	Radon	^{219}Rn	α 6.4, 6.8	4.0 s
Actinium A ~100% | ~5 X 10^{-4}%	Polonium	^{215}Po	α 7.4	1.83 X 10^{-3} s
Actinium B ↓ Astatine-215	Lead Astatine	^{211}Pb ^{215}At	β 1.4, 0.5 α	36.1 m 10^{-4} s
Actinium C 99.7% | $0^{.3}$% Actinium C"	Bismuth*** Polonium	^{211}Bi ^{211}Po	α 6.3, 6.6 α	2.15 m 0.52 s
Actinium C" ↓	Thallium	^{207}Tl	β 1.44	4.79 m
Actinium D (End Product)	Lead	^{207}Pb	Stable	

SUBJECT INDEX

Accumulator, 50, 51
Actinium, 29, 32
Actinium series, 12-13, 134
Actinon, 13, 25, 32, 40
Adsorption method, 52
Aerosols
   concentration of, 87, 117
   continental and marine, 61, 86
   decay products, 85
   diurnal variation, 92-93
   indoor and outdoor
      environments, 87
   number concentration, 87
   size-number distributions, 86
   sources and size, 85
   vertical distribution, 88
Age determination, 20
Alpha decay, 15
Alpha particle
   in living tissue, 126
   range in air, 126
Atomic number, 14-15
Atmospheric radon, 59
   continental and marine air
      masses, 61, 86
   diurnal variation, 62, 65
   ground level, 59
   seasonal variation, 66
   vertical distribution, 60
Beta decay, 15
Beta particles, 126
Books on radon, 129
Carlsbad Caverns, 49, 92, 99
Charcoal absorption, 26
Chemical activity, 26-27
Chemistry, 25-26
Commission of the European
   Communities, 128
Conductivity, 94
Continental air masses, 61
Control methods, 118
Cosmic radiation, 7
Cumulus convection, 75
Curie, Pierre and Marie, 2-3, 6
Decay series, 12-13, 132-134

Deposition velocity, 117
Diffusion and viscous flow,
   44-47
Diffusion length, 47
Diffusive flow, 44
Discovery of radon, 2
Diurnal and seasonal scales, 62
Diurnal changes, 63-65
Diurnal variation, 63
Diurnal variation of positive
   ions, 93
Dose equivalent (mSv/y), 7, 127
Drift velocity, 83, 117
Effective dose, 7, 127
Effective ^{226}Ra, 113
Electrical character of
   atmosphere, 91
   conductivity, 94
   ion concentration, 92
   ionization, 91
   mobility, 94
Emanation properties,
   35-36, 55
Environmental Protection
   Agency, U.S., 121
Equilibrium Equivalent Concen-
   tration (EEC), 126-127
Equilibrium and plateout, 116
Flow in channels, 48
Flow method, 52
Fluorescence, 5
Gamma rays, 4, 126
Health effects, 125-130
Historical perspective, 1
Human dose, 127
   aerosol concentrations, 127
   bronchial tissue, 127-128
   ions and molecular
      clusters, 127
   respiratory system, 127-128
Impact, 128-129
Indoor radon, 113-124
   construction materials,
      114-115
   equilibrium fraction, 116

methods for control, 118
natural gas, 115
plateout, 117
soil, 113, 118
sources, 113-116
surveys, 119-122
water supplies, 114
ventilation, 117
Inhaled radiation, 7
Ion density, 92-93
Ionization, 91-92
Isotopes of radon, 25
Diffusion in fine
 capillaries, 46
Lung cancer, 6, 104
Marine air masses, 61
Mines, 101-104
Mines other than uranium,
 105-106
Mobility of the ^{222}Rn decay
 product ions, 83, 94
Molecular diffusion, 47
National Council on Radiation
 Protection (NCRP), 129
Natural radioactivity decay
 series, 12-13, 132-134
Neutrino, 15
Nocturnal air drainage, 74
Nobel Prize, 6
Nuclear structure, 13, 15
Outdoor radon, 113
Periodic table, 13-14
Physical properties (^{222}Rn),
 26-27
Pitchblende, 2, 101
Plateout, 117
 deposition velocity, 117
Poiseuille flow, 47
Potential alpha energy (PAEC),
 126-127
Radiation environment, 6
Radiation from radon and its
 decay products, 125
 exposure and dose definitions,
  127-128
 natural decay series in the
  air, 12-13

quantities and units,
 126-127
Radioactive
 decay, 9, 15, 17
 equilibrium, 18
 growth, 17
Radioactivity, 9
Radium, 32, 55
 crustal abundance, 33, 55
 emanation, 35, 37
 isotopes, 13, 34, 36
Radon-219 (actinon), 40
Radon-220 (thoron), 36, 38-39
 76-77
Radon-222, 37-39
 discovery of, 2
 diurnal variation, 63
 in houses, 113-124
 outdoors, 63-66, 113
 physical properties, 13, 26
 seasonal variation, 66-69
 soil gas, 73
 soil to air, 43
 solubility, 27
 sources, 29, 36-39
 transport, 43-49
 water supplies, 114
Radon decay products in the
 atmosphere, 81
 mean life, 82
 mobility of ions, 83
 tracers in the environ-
  ment, 84
 unattached, 82
Radon flux density measurement
 accumulator, 50
 adsorption method, 52
 flow method, 52
 soil concentration gradient, 54
 vertical profile, 53
Radon in the atmosphere, 59
Radon in water
 public water supplies, 109
 sea water, 110
Radon isotopes, 36-40

Radon underground, 97
　　mines, 101, 104
　　mines other than uranium, 105-106
　　tunnel, 99-101
　　underground cavity, 98
　　uranium mining, 102-105
Risk from radon in water, 109-119
Rutherford, Ernest, 3-5
Sealing, 118
Seasonal variation, 66-68
Sea water, 110
Secular equilibrium, 18, 21
Short-lived decay products, 38, 72
Soil concentration gradient, 54
Soil, 35, 113, 118
Sources of radon, 29, 113-115
　　construction materials, 114
　　natural gas, 115
　　soil, 118
　　water supplies, 114
Spas and therapeutic galleries 129-130
Surveys, 119-120
　　Cohen, B.L., 121
　　international, 120
　　Nero, A.V., 121
　　New Mexico, 122
　　of indoor radon, 120
　　United States, 121
Tables of radioactive isotopes, 13, 132-134
Terrestrial radiation, 6-7
Thorium, 29, 31
Thorium series, 13, 133

Thoron, 39
　　physical properties, 39
Thunderstorms, study of, 84
Tracer in the atmosphere, 71-77
Tracers in the electrical environment, 84
Transport processes, 43, 71
　　diffusion and viscous flow 44-45
　　flow in channels, 48, 99
Transient equilibrium, 19, 21
Turbulent diffusion, 8, 73-74, 77
Unattached radon decay products, 81
Uranium, 29
Uranium-238 and thorium-232 concentrations, 30-31
Uranium mines, 103-105
Uranium series, 13, 132
Ventilation, 117
Vertical convective transport, 60, 73-74
Vertical distribution, 53, 60
Vertical distribution of aerosols, 88
Vertical profile, 53, 60
Viscous flow, 44
Volcanoes, 49
Water supplies
　　aquifers, 109
　　domestic water, 109
　　rain water, 109
　　sea water, 110
　　surface water, 109
Working Level (WL), 126
X-rays, 4, 6